Signals and Communication Technology

More information about this series at http://www.springer.com/series/4748

Jaco du Preez · Saurabh Sinha

Millimeter-Wave Antennas: Configurations and Applications

Springer

Jaco du Preez
University of Johannesburg
Johannesburg
South Africa

Saurabh Sinha
University of Johannesburg
Johannesburg
South Africa

ISSN 1860-4862   ISSN 1860-4870   (electronic)
Signals and Communication Technology
ISBN 978-3-319-81714-9   ISBN 978-3-319-35068-4   (eBook)
DOI 10.1007/978-3-319-35068-4

Printed on acid-free paper

This Springer imprint is published by Springer Nature
The registered company is Springer International Publishing AG Switzerland

Preface

Since the inception of wireless systems, engineers have continuously strived to develop systems that utilize higher frequencies, in pursuit of numerous advantages that it may offer to particular applications. As a result, the components that constitute a typical wireless system—that is, the antenna and the analog and digital front ends—have been fiercely studied and remarkable developments have surfaced in the last century. While millimeter-wave systems are relatively new (only experiencing any real development since the 70s), they are arguably one of the fastest growing fields in wireless research. Rapidly developing standards and component technology have two major driving forces behind the tremendous growth that we have come to witness in the past decade.

For researchers and designers entering this field, the vast extent of information, research results and systems that exist can be gruesomely overwhelming, which is where this text is intended to assist. This text is heavily focused on current research, and it is designed to cover as many different antenna configurations and applications as possible. It should therefore enable a designer (or researcher, for that matter) to make an informed decision on one of the most critical components of a wireless system. As the text progresses, the reader will notice that special attention is paid to measurement techniques and practical results. While fundamental concepts form the baseline of each system that we intend to discuss, there are a multitude of other books that intensely cover relevant theory. Nonetheless, where it is deemed applicable, this text does attempt to cover fundamental theory, but it is assumed that the reader has sound understanding of wireless systems and antenna concepts.

Discussions in the introductory chapter are intended to provide an idea of development trends witnessed in millimeter-wave systems, and stretches from the earliest stage of the field up until modern times, leaving the reader ready to expand into the next six chapters. A second key part of the introductory chapter is a discussion on spectrum management and allocation, which is the first and perhaps the most crucial component necessary to develop any wireless system. Following this, there is no prescribed order in which the next five chapters should be approached, although there are minor references between said chapters, it is written so that the reader never feels disconnected from the core concept of the text.

Chapters 2–5 focus on typical antenna types that are scalable to millimeter-wave dimensions. We begin with leaky-wave and surface-wave antennas in Chap. 2, a fundamental branch of millimeter-wave antennas, and one whose principles resonate through the remainder of the book. We cover some basic design principles and concepts unique to leaky-wave antennas, which can be synthesized from a number of different structures, such as rectangular waveguides and microstrip arrays. Various types of printed antennas are discussed in Chap. 3, where we once again spend a short amount of time to cover some fundamental theory, before delving into recently publications on the millimeter-wave scaling of these antennas. Printed antennas are highly desirable in applications with severe space constraints, and these antennas are typically capable of mechanical self-support. To cope with the extreme mechanical tolerances at such high frequencies, several configurations have been developed that, in some ways, mitigate the implications of these tolerances, along with changes in approach to fabrication.

In Chap. 4, one of the most actively developed fields at the moment, integrated antennas, is discussed. The requirements of developing radio standards in the 60, 77, and 94 GHz bands—compact size, low power, and the ability to mass produce—are truly well catered for by integrated solutions where the antenna is fabricated on the same substrate as supporting circuitry. As a result, a strong emphasis is placed on 60 GHz radio standards and the accompanying developments. Chapter 5 is the final one where the antenna is the focal point, discussing millimeter-wave reflector and lens systems. Highlights of this chapter include the shift away from traditional reflector configurations (such as the popular parabolic dish) toward more compact structures such as reflectarrays and lenses. Furthermore, we discuss beam forming systems implemented with lens antennas and their many potential advantages in the domain of short-range wireless links.

The last two chapters are intended to tackle the antenna question from a systems perspective, overviewing supporting circuitry and providing a detailed look into millimeter-wave applications. In conclusion, based on numerous discussions on manufacturing, design principles, and practical results from leading research groups, this text should aid designers and researchers to better understand their application and its required antenna system.

The authors would like to recognize the research capacity grant of the Dept.: Higher Education and Training, South Africa for sponsoring this work. Furthermore, the authors would like to recognize Dr. Riëtte de Lange, Postgraduate School, University of Johannesburg, South Africa, for her effective administration of this grant.

Johannesburg, South Africa   Jaco du Preez
                                                                   Saurabh Sinha

Contents

Contents

Abstract

The text on *Millimeter-Wave Antenna Technology* brings forward the state of the art in antennas designed for millimeter-wave applications. The spectrum from 30 to 300 GHz is recognized as millimeter-wave, and frequency bands at 60 and 77 GHz are of particular interest, with major applications such as short-range communications and automotive radar.

The drive toward 60 GHz wireless communications is a result of the major potential benefits offered by implementing systems in this band, two of which are miniaturization and a significant improvement on data rates compared to the 2.4 and 5 GHz systems used today. Aside from commercial applications, millimeter-wave systems are beneficial in defense applications as well, with short-range radars and communication networks being two areas of interest. Given the development of new integrated circuit technologies, miniaturized nodes, there are newer alternatives to implement mm-Wave systems on-chip and on-package. These implementation methodologies include the possibility for on-package antennas.

This text aims to provide a comprehensive research results on recent advances in this field, and several popular types of antennas are considered. Peripheral circuitry such as switches and feed networks are mentioned, and some of the challenges related to millimeter-wave design are discussed.

Chapter 1
Introduction to the World of Millimeter-Wave Systems

We begin the investigation into the fascinating world of millimeter-wave antennas with a brief historical overview of millimeter-wave development. It is not intended as a comprehensive historical reference, the aim is to provide context on the matter and highlight major advances in the field over the past few decades. Several key publications are cited throughout this chapter, which should provide a more detailed perspective. Expanded development of millimeter-wave systems in recent years can largely be attributed to significant improvements in component technology, newly formulated spectrum allocations, and consistent improvement in manufacturing techniques.

One of the most critical considerations in modern electronics development is the electromagnetic spectrum and its efficient utilization. With an ever-increasing number of devices being dependent on sufficient spectrum access, managing and distributing the spectrum among application areas is critical. Millimeter-wave systems are advantageous in this regard, and their place in the electromagnetic spectrum is discussed and followed by an overview of wave propagation mechanics at 30 GHz and upwards.

1.1 A Brief History of Millimeter-Wave Electronics

Millimeter-wave antennas have enjoyed an extended history of development as these systems as a whole have evolved. Being able to provide high spatial resolutions with small physical dimensions, millimeter-wave antennas are crucial in the development of millimeter systems [1]. Millimeter and submillimeter waves have been investigated since the earliest stages of electromagnetic theory. The history of the development of millimeter-wave technology can be divided into three periods: From the discovery of electromagnetic waves until the end of World War II,

© Springer International Publishing Switzerland 2016
J. du Preez and S. Sinha, *Millimeter-Wave Antennas: Configurations and Applications*, Signals and Communication Technology,
DOI 10.1007/978-3-319-35068-4_1

Fig. 1.1 Spark-gap
transmitter circuit

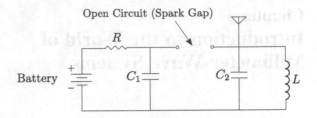

the postwar period of 1946–1965, and the modern era of millimeter-wave technology after 1965 [2]. At the time that Hertz experimentally proved Maxwell's
theory on electromagnetic waves in the early 1890s, he managed to generate radio
waves with a spark-gap transmitter. These waves were measured to be roughly in
the centimeter range [3]. A typical spark-gap transmitter circuit is shown in
Fig. 1.1.

Other scientists, such as Lebedew and Lampa, followed on Hertz's work by
detecting wavelengths of 6 and 4 mm, respectively. These results were, however,
greeted with uncertainty by the scientific community at the time and in 1923
Nichols and Tear stated that the wavelengths produced by Lebedew and Lampa had
probably been underestimated.

J.C. Bose (shown in the photograph in Fig. 1.2) is believed to be the first
researcher to conduct quantitative measurements down to 5 mm [4]. Bose, inspired

Fig. 1.2 Jagadish Chandra
Bose photographed with his
demonstration of a horn
antenna

by the work of Hertz, was a pioneer in radio science and is credited with several remarkable discoveries. These include—among many others—his millimeter-wave spark-gap transmitter Galena (PbS) detector sensitive to millimeter waves, dielectric lens, millimeter wave link for remote control, wire grid polarizer and horn antenna.

As an alternative to the Hertzian spark-gap source, Rubens and his colleagues were working with an infrared heat source and so-called "reststrahlen" (translating to "residual rays") interference filters, which served to generate quasi-monochromatic beams of radiation. Producing over 140 papers between 1892 and 1922, this period has often been referred to as the "Rubens era" of infrared research, and rightfully so. Nichols and Tear systematically closed the gap between radio frequency (RF) and optical regions in the 1920s. Their modified version of the Hertzian spark-gap transmitter utilized a reflecting echelon grating in order to facilitate wavelength measurements. Constant improvements on their system eventually led to measured wavelengths of down to 0.22 mm, which were followed by reports from Russian researcher Glagolewa-Arkadiewa, who managed to generate as well as detect wavelengths from 50 mm all the way down to 0.082 mm. The popular spark-gap transmitter was, however, not a very stable signal source and did not bear a close resemblance to a coherent oscillator. Several new types of coherent RF sources were being developed in the 1930s. Most of these were built from different configurations of vacuum tubes. Two of these were known as the klystron and the cavity magnetron, and these devices played a crucial role in radar development during World War II.

The klystron is a specialized type of linear particle accelerator. Its invention is credited to the Varian brothers and it was formally announced in the February 1939 issue of The Journal of Applied Physics [5, 6]. One of the early klystrons at Stanford was used in an experimental Doppler radar. A similar system was completed in South Africa in December 1939 and tested at the University of the Witwatersrand. The earliest tests yielded successful detections of up to 10 km [7].

The cavity magnetron (shown in Fig. 1.3), invented in 1939 by Randall and Boot at the University of Birmingham, was an improvement on the earlier split-anode magnetron, which was incapable of generating microwave frequencies [8]. However, neither the klystron nor the cavity magnetron was capable of generating signals in the millimeter-wave region, predominantly because most airborne radars at the time operated successfully in the X-band (8–12 GHz). As the war continued, experiments were conducted on devices that operated in the K-band (12–40 GHz) and upwards. This led to the important discovery that water vapor absorbs large portions of electromagnetic radiation at 22.3 GHz, which in turn led to the well-known upper and lower K-band (known as the K_a and K_u bands, respectively) distinction [9].

World War II brought about another important discovery in the traveling wave tube: a wideband, high-power, microwave amplifier [10, 11] (shown in Fig. 1.4).

With a host of stable, coherent microwave sources available, the postwar years saw widespread development of microwave and millimeter-wave spectroscopy. Walter Gordy, a renowned physicist in the field of microwave spectroscopy, joined Duke University in 1946. He proceeded to institute a research group focused on

Fig. 1.3 The first microwave cavity used in the Randall and Boot's cavity magnetron

Fig. 1.4 Diagram of a traveling-wave tube amplifier

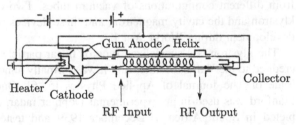

microwave and millimeter-wave spectroscopy and his students produced many high-resolution spectral measurements. Many of the students who graduated under his guidance participated in millimeter-wave programs at other organizations, most notably the Bell Telephone Labs and Johns Hopkins University. In fact, the Bell Labs had initiated millimeter-wave research prior to World War II, with early work on waveguides by Southworth [12] and Schelkunoff [13] being greatly influential.

The interest in using millimeter waves for communication systems grew steadily as its advantages became more apparent. Small antenna apertures capable of producing high gains and components offering bandwidths in the gigahertz range were two of the most attractive possibilities. However, the atmospheric absorption of electromagnetic waves had not really been investigated until Beringer published his measurements on oxygen absorption at 60 GHz in 1946 [14]. In the years following Beringer's work, an analysis from van Vleck in 1947 and a detailed set of measurements taken by Gordy's group in 1949 were published [15]. While the atmospheric attenuation suffered by millimeter-wave signals is a major disadvantage when considering conventional longer range communication systems, its potential advantage in short-range applications was not completely understood until much later.

The research at Bell Labs, producing many new sources, components and millimeter-wave detectors, climaxed in 1970 upon completion of their millimeter waveguide-based communication system in New Jersey [16], which covered a distance of 14 km.

As the research efforts from the radio end of the spectrum progressively grew in frequency in the 1950s, so too did the research at the infrared and far-infrared end approach the millimeter-wave region. Considerable growth in interest arose in millimeter-wave research as more and more institutions became involved in the field in the late 1950s. The group at the Georgia Institute of Technology expended great efforts toward antennas and radar systems, and a 70 GHz radar was eventually developed into a production system [17]. Scattering phenomena in the millimeter radar bands, as well as the development of high-performance antennas at 70 GHz and 95 GHz, were also researched at Georgia Institute of Technology during this time. Radar work continued steadily and in the early 1980s a 225 GHz radar became operational (the highest frequency radar in the world at the time, shown in Fig. 1.5).

The mid-1950s brought about higher frequency versions of the well-known klystron and magnetron signal generators. At the forefront of magnetron development was the Columbia Radiation Laboratory [18]. The group focused on shortening the wavelengths and increasing the power output from generated signals, as well as developing wideband tunable tubes. Another interesting development was the so-called low field operation of the magnetron [19].

In the Netherlands, the klystrons developed at the Phillips Research Laboratories reached frequencies of up to 120 GHz by 1959 [2], as Bell Labs were developing

Fig. 1.5 Engineer at GTRI adjusting the 225 GHz radar system in 1981. Photograph courtesy of the Georgia Tech Historical Archive

traveling-wave tube amplifiers that operated at up to 58 GHz. The maiden issue of the IRE Transactions on Microwave Theory and Techniques entered print in 1953, and a year later an entire issue of this publication was dedicated to millimeter-wave topics. Much emphasis was placed on new stable sources, mixer configurations and the possibilities of new transmission lines. Conventional rectangular waveguides (such as the one shown in Fig. 1.6) are reasonably lossy around 100 GHz (around 1–10 dB/km [20]), and an alternative circular structure that supported the TE_{01} mode was proposed and developed [21].

The helix waveguide structure was primarily developed at Bell Labs, but did not lend itself well to fabrication [22]. King reported a number of components that could be used with a new type of transmission line (called the dielectric image line), which was based on the symmetry of a dipole mode in a dielectric rod [23]. Numerous measurements at K-band were carried out and published by King [24, 25]. The conducting image plane does, however, contribute to the attenuation levels of the dielectric image, but it is nonetheless below that of a rectangular waveguide. A detailed characterization of the dielectric image line and components associated with it was extended further into the millimeter-wave region by Wiltse. His measurements and analyses were performed at around 35–140 GHz [26].

Wiltse and his colleagues also performed thorough analyses on the propagation of surface waves on wires (both dielectrically coated and uncoated). They proposed the phase-corrected Fresnel zone plate as an alternative to the familiar dielectric lenses and prototypes of these plates were designed and constructed to use at 140 and 210 GHz [27].

The low-loss H-guide transmission line received substantial attention after it had been proposed by Tischer in the 1950s. Tischer extensively studied properties of both single and double slab variants of the H-guide at X-band and millimeter-wave frequencies [28]. Cohn expressed some concerns about Tischer's work in the sense that the results obtained from his own analyses differed from those of Tischer. He later explained that his approximation was based on the field distribution in a lossless waveguide, and although it led to the same end result obtained by Cohn, the two were still in disagreement on the methodology [29]. Not long after, a Russian scientist by the name of Guttsayt published his own findings on the H-guide and in comparing his results for the attenuation factor with those obtained by Cohn and Tischer, he argued that both their results were valid [30]. An illustration of a shielded helix waveguide is given in Fig. 1.7. As an increasing number of papers on

Fig. 1.6 Structure of a rectangular waveguide, filled with a dielectric medium

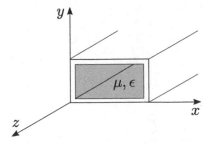

Fig. 1.7 A shielded helix waveguide

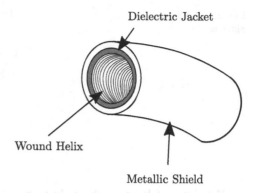

Dielectric Jacket

Wound Helix

Metallic Shield

millimeter-wave electronic components started appearing, Sharpless published his wafer-type silicon point-contact rectifier [31]. This attracted considerable attention, as it was a significant improvement on the packaged diodes available at the time, since these were very difficult to tune and could not operate over wide bandwidths. Another benefit was reproducibility, and Sharpless emphasized this by performing measurements of hundreds of rectifiers and evaluating their consistency. His units were shown to have no measureable deterioration over several months of laboratory testing, offered good impedance matching and could be manufactured with reliable consistency.

Increased sensitivity radiometers were introduced early in the 1960s and these operated at a much higher intermediate frequency (IF) (in the gigahertz range) to circumvent the problem of oscillator noise [32]. This effectively removed the requirement of a balanced mixer and thus made it possible to realistically utilize higher IF bandwidths, leading to significantly increased thermal sensitivity. These radiometers operated at 140 and 225 GHz and the use of single-ended mixer configurations together with a wide IF bandwidth is commonplace in microwave systems today.

Around this time, early high-gain reflectors were being demonstrated at the Aerospace Corporation. One particular antenna (which has been operational since 1963) is usable up to 220 GHz with a measured gain of 70.5 dB at 94 GHz, roughly corresponding to an aperture efficiency of 55 % [33]. Another high-gain structure is the lens antenna, and perhaps the earliest millimeter-wave lens was demonstrated by Luneburg in the late 1950s [34]. This antenna, using a two-dimensional structure with contoured parallel plates, produces a fan beam, which then illuminates a parabolic reflector. The first demonstrations were around 70 GHz and the structure supported both azimuth and elevation steering. A multiple beam feed facilitated azimuth steering, while the positioning of the reflector could be varied in order to realize elevation steering.

Deriving microwave antennas from the Cassegrain telescope was first reported in 1961 by Hannan [35]. The idea was fairly simple and it introduced a new type of reflector configuration (referred to as the Cassegrain antenna, shown in Fig. 1.8) that lends itself very well to microwave and millimeter-wave designs.

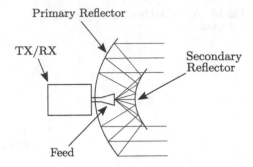

Fig. 1.8 A Cassegrain antenna

The dawn of lasers in the early to mid-1960s unfortunately diverted a lot of attention from the research community away from development in the millimeter and submillimeter regions. The inevitable result was that large amounts of funding were redirected toward laser research, as it was a new and exciting technology that offered numerous practical advantages. However, developments of significant importance still took place in the 1960s. In 1962, the Gunn effect was discovered in n-type GaAs; it was demonstrated that current oscillations can arise at microwave frequencies [36].

There were many possible explanations for why these oscillations were observed (many of which were rejected outright by Gunn), and Knight and Peterson quantified many of the phenomena related to this effect [37]. Shortly afterwards, the IMPATT oscillator was developed at Bell Labs [38] and by the 1970s, both of these oscillator configurations could be obtained commercially and operated at frequencies of up to 35 GHz (although the early experimentation in pulsed and CW modes was performed up to 84 GHz). Several improvements on the IMPATT oscillator were published in the following years, primarily focused on improving the efficiency and noise performance of the system.

As the quest for solid-state devices with long lifetimes and high reliability continued, several programs were initiated in the U.S. and England. Particular interest was shown by the U.S. Army Ballistics Research Laboratory in millimeter-wave components used in defense programs. Similar defense-oriented programs were running at the Royal Radar Establishment in England.

Led by Duke and BenDaniel at General Electric Research, the initial publication on the metal-oxide-metal diode appeared in 1966. This was followed by a myriad of articles in the years that followed [39]. Undoubtedly, the arrival of stable solid-state sources was a critical enabler for millimeter-wave systems. The July 1970 issue of the IEEE Transactions on Antennas and Propagation was dedicated to millimeter-wave systems, and the technical progress of millimeter-wave research continued to expand steadily throughout the 1970s. The interdependence of components and sensors is highlighted in Fig. 1.9.

Integrated antenna systems gained momentum in the 1970s and 1980s, as integrated circuit technology grew more sophisticated and many groups were investigating and experimenting with millimeter-wave communication links.

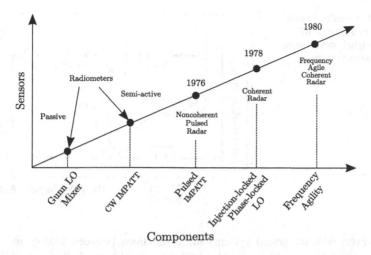

Fig. 1.9 Development of millimeter-wave sensor technology and solid-state circuits

A V-band dielectric rod was reported by Shiau in 1976, and the final antenna (several configurations were experimented with) was integrated into V-band transmitter and receiver modules [40]. Wiltse and Black published their paper on Fresnel zone plates in 1987 with emphasis on the performance of the antenna at millimeter wavelengths [41]. Zone plates are simpler to construct and provide a lower absorption loss level, as well as smaller physical dimensions when compared to lenses, making them highly desirable when space is a limiting factor. A cross-sectional comparison of plates and lenses is shown in Fig. 1.10.

With solid-state circuits becoming increasingly sophisticated, system-on-chip solutions rely heavily on the performance of the antenna system. The architecture of a generic RF system is shown in Fig. 1.11. It aims to highlight the changing trend toward integrated systems. Traditionally, each subsystem is designed relatively independently from the others (aside from interfacing and packaging specifications), granting the designers for each section a lot of freedom to select the best materials and components.

Fig. 1.10 Comparison of lenses and zone plate structures

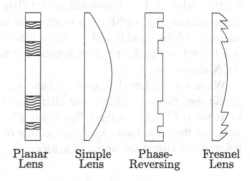

Planar Lens   Simple Lens   Phase-Reversing   Fresnel Lens

Fig. 1.11 Simplified block
diagram of a generic RF
system: digital, mixed signal,
RF and antenna sections

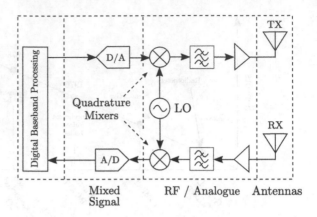

Mixed   RF / Analogue   Antennas
Signal

However, with integrated systems the distinctions between analog and mixed
signal sections become blurred and introduce several new challenges, while providing substantial benefits in terms of physical size.

As the components that support millimeter-wave communications, sensing and
control applications continue to mature, spectrum licensing and standardization are
equally important.

1.2   The Millimeter-Wave Spectrum

The deciding factor in a high data rate in wireless systems is spectrum. With
increasing data transfer speeds, the bandwidth requirements become proportionally
larger. By definition, electromagnetic waves in the millimeter-wave region are those
with wavelengths that stretch from 10 to 1 mm. In terms of free-space propagation,
these wavelengths correspond to frequencies of 30–300 GHz. While the data rate in
a communication system is a function of bandwidth and symbol rate, a larger
bandwidth will almost always lead to an increase in data transfer speed. This simply
means that communication systems that operate on higher frequencies are capable
of producing higher data rates, seeing that a 1 % bandwidth at 600 MHz is equal to
6 MHz, while a 1 % bandwidth at 60 GHz is equal to 600 MHz. Using binary
phase shift keying (BPSK) as a modulation scheme, 600 MHz translates into a data
rate of 600 Mbps, and 6 MHz equals a data rate of 6 Mbps [42]. Shown in Fig. 1.12
is a breakdown of the electromagnetic spectrum, from high frequency (HF) up to
the X-ray region.

While the preferred frequency band nomenclature differs for just about any
application, the convention most often used is radar frequency bands (also referred
to as the IEEE radio bands). Spectrum allocation differs from one country to the
next, but there are many applications that overlap. Most of these will be found in
government issued spectrum allocations.

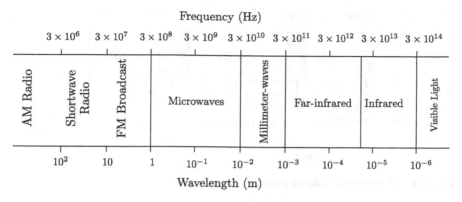

Fig. 1.12 The electromagnetic spectrum

As mentioned on numerous occasions, the frequency range of 30–300 GHz is of particular interest and the recent developments regarding spectrum allocation in this frequency band is at the center of this discussion. Perhaps the most familiar millimeter-wave band is centered at 60 GHz. In the US the 60 GHz band is allocated to short-range, unlicensed communication links (and has been since 2001 [43]) and it is the specified frequency for the IEEE 802.11ad standard.

As the standard matures and increasingly sophisticated technology becomes available that serves to facilitate 60 GHz networks, communication authorities from various countries allocate spectrum around this band. The current allocation for a handful of countries is shown in Fig. 1.13.

Given the existing congestion at cellular bands (generally in the range of 800/2100 MHz, depending on the region), widespread implementation of short-range communication links in the 60 GHz band could relieve a significant amount of spectrum overcrowding that exists today. Consider the diagram in Fig. 1.14, which shows spectrum usage in the bands above 3 GHz.

For most of the modern era of radio communications the vast majority of systems have been (and still are) designed to operate in the narrow band of 300 MHz–3 GHz. Examples of these include cellular networks, Wi-Fi, GPS, and radio broadcasting. In terms of propagation characteristics, this is an extremely favorable frequency band to work with, especially in commercial wireless applications. It is only in recent years that communications authorities have seriously begun to investigate millimeter bands as a solution for mobile networks.

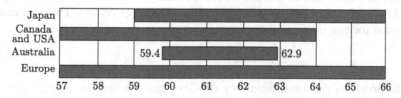

Fig. 1.13 60 GHz Spectrum allocation

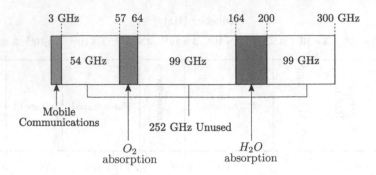

Fig. 1.14 Millimeter-wave frequency bands

The microwave bands between 6 GHz and 40 GHz are typically used for cellular backhaul and long-distance links, providing data rates that are in the range of several hundred Mb/s. To extend data rates into the Gb/s range, utilizing the spectrum at 60 GHz and higher is crucial. While a large portion of the spectrum discussion has been focused on the 60 GHz band, this is by far not the only millimeter band of interest. A well-known utilization of the 77 GHz band lies in automotive radar and vehicle safety systems. In Europe, the 77–81 GHz window is officially allocated to ultra-wideband short-range radar. Millimeter wavelengths allow for sophisticated radar antennas to be integrated with silicon circuits, greatly reducing the physical size of the system. The 77 GHz band has also been used for microwave imaging and short-range surveillance networks. Other notable bands are at 35, 94, 140, and 220 GHz, supporting a wide range of applications from radio astronomy to security screening devices.

Often synonymous with a discussion on the electromagnetic spectrum, is the topic of wave propagation at higher frequencies.

1.3   Wave Propagation at Millimeter-Wave Frequencies

The propagation mechanics concerning electromagnetic waves have been intensely studied since 1901, when Marconi demonstrated that it is possible to transmit signals across the Atlantic Ocean. Essentially, four mechanisms account for over-the-horizon propagation of signals [44]: diffraction, reflection, refraction and transmission by means of surface waves. Primarily, the transmission loss experienced by microwave systems is accounted for by free space loss. This loss can be computed for two isotropic antennas as

$$FSL_{dB} = 92.4 + 20\log f + 20\log R, \tag{1.1}$$

where f indicates the signal frequency (in GHz) and R is the line-of-sight distance between the transmitting and receiving antennas (in km).

As the system frequency increases into the millimeter-wave region, several other influential factors become apparent. A common misconception is that propagation loss is frequency-dependent, meaning that signals at higher frequencies cannot propagate as well as if they were transmitted at a lower frequency. This is not always the case, seeing that a greater number of antennas can be packed into the same physical space at shorter wavelengths, increasing the effective aperture [45]. A significant increase in the number of antennas in millimeter arrays also results in highly directive steerable beams.

Typical sources are absorption by atmospheric gases (such as water vapor and oxygen), diffraction, precipitation attenuation, blockage by foliage and scattering effects [46]. Range attenuation drastically increases when several of these effects are present. The approximate variation in attenuation that electromagnetic radiation experiences over frequency is depicted in Fig. 1.15.

Figure 1.15 highlights the intense oxygen absorption suffered at 60 GHz and the reasoning behind its preference for short-range radio links.

Furthermore, the loss experienced as a result of reflection and diffraction largely depends on the surface and material, and it is most prominent in systems where there is no direct line of sight between the transmitting and receiving antennas. Millimeter-wave signals, unlike their lower frequency counterparts, generally cannot penetrate solid materials without suffering significant losses.

Listed in Table 1.1 are typical attenuation values for signals propagating through common materials [45].

Losses suffered from foliage blockage can be substantial at millimeter bands, and based on empirical data, the loss through 10 m of foliage can be as high as 23.5 dB at 80 GHz [45]. In terms of precipitation (such as rain or snow), wavelengths are similar to the size of raindrops, leading to scattering of the RF signal. The attenuation suffered from rainfall can be derived from the rainfall intensity in millimeters per hour [47]. Light showers (2.5 mm/h) cause attenuation of approximately 1 dB/km, while propagation in heavy rain (25 mm/h) results in an attenuation of 10 dB/km at millimeter wavelengths. This can be a severe disadvantage in any type of outdoor communication system.

Multiple propagation paths produce multipath interference. Each path is associated with specific attenuation and time delays, which arise primarily from surface

Fig. 1.15 Atmospheric attenuation experienced by electromagnetic waves at sea level

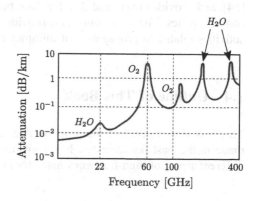

Table 1.1 Material attenuation (in dB) experienced by 3, 40 and 60 GHz signals

Material	Thickness (mm)	3 GHz	40 GHz	60 GHz
Drywall	25	5.4	–	6.0
Concrete	100	17.7	175	–
Clear glass	40	6.4	2.5	3.6
Wood	70	5.4	3.5	–
Mesh glass	100	7.7	–	10.2

reflections and scattering. Scattering occurs when the scattering object is physically similar to the signal wavelength, and with signals exhibiting limited penetration through objects, the effects of these phenomena are diminished. However, reflections become an issue, since they arise from objects that are physically larger than the signal wavelength and a multitude of objects thus become reflectors at millimeter wavelengths [48]. The temporal spreading effect of a wireless channel is quantified by the root mean square (RMS) delay spread. Typically, if a wireless channel were to experience a delay spread of T_{RMS}, then it will suffer inter-symbol interference of $|f_s T_{RMS} - 1|$, provided the symbol frequency is given by f_s Considering the fact that 60 GHz systems typically function on very wide bandwidths, even minute values of T_{RMS} can result in severe inter-symbol interference. Detailed measurements on the RMS delay spread observed in wireless channels have been reported numerous times [49].

Careful selection of transmit and receive antennas can alleviate the multipath issue to an extent. For example, increasing the directivity would serve to limit the angular range from which the antenna can receive (or transmit) RF energy. Another solution is to opt for circular polarization (as opposed to linear), as this has been shown to reduce RMS delay spread values by a factor of 2 and upwards [50].

Experimental and theoretical results on propagation phenomena have been extensively reported, focusing on satellite communications [51], control and communication networks in railway trains [52] and mobile links in dense urban environments [53]. A comprehensive analysis of the reflection characteristics exhibited by signals in millimeter-wave bands has been performed and reported [54], and provides measured data for floor boards, ceiling boards, tile carpets, and concrete plates. These measurements provide fundamental information required for activities related to propagation of millimeter waves.

1.4   Outline of This Book

This text aims to outline the progress that millimeter-wave antenna research has made in the past few decades. It is split into seven chapters, each focusing on a different aspect of high-frequency antennas in the hope of providing comprehensive insight into the matter. This introductory chapter provided, from a historical viewpoint, the early stages of millimeter-wave development and it covered some basic principles concerning the electromagnetic spectrum from 30 to 300 GHz. For the remainder of this book, each chapter follows a similar approach. Each subset of antennas discussed here can be further subdivided, based on several key differences. As an illustrative example, printed antennas can be dipoles, microstrip patches or any number of other shapes.

Following this, each chapter will provide a short overview of the theory of operation of each subset of antennas. However, this is not intended as the sole source of a theoretical background on the matter and the reader is, as always, encouraged to view the cited works for greater detail. Followed by the theoretical overview, mentioning the design principles should provide the reader with a basic understanding of the design process. With the context established, the review of current literature then follows.

Most guiding structures used at millimeter waves, such as microstrip lines and dielectric guides, are open. The effect of this is that energy leakage occurs in cases where the guide is not excited in the appropriate mode, or perhaps when discontinuities exist in the guiding structure. These antennas are commonly known as leaky wave antennas, which are derived from open and closed millimeter waveguides, and they are the focal point of Chap. 2.

Printed antennas are structurally simple and it is generally possible to fabricate them through photolithographic methods. Their lightweight, low-profile nature allows these antennas to be expanded into large array configurations, well suited to integrated circuit applications. However, their extension into millimeter regions is not a simple case of wavelength scaling, as extreme manufacturing tolerances (among other factors) come into play. Thus Chap. 3 details the numerous design issues concerning planar antennas, as well as noteworthy designs and integrations into functional RF systems.

Most modern wireless systems comprise four functional units: digital baseband processing, an interface between digital and analog circuitry, an RF front end to provide a modulated carrier signal and an antenna to facilitate free-space transmission and reception of signals. Originally, these four functional systems were designed as distinct systems that can later be integrated. This is a sensible approach, seeing that digital systems are well suited for CMOS technology, while it is preferable to implement RF circuitry and antennas on low-loss printed circuit boards, such as Rogers, Duroid and FR-4. However, given the size and weight constraints of modernized systems, the trend is starting to shift toward system-in-package and system-on-chip solutions. Integrated antennas are arguably the last remaining bottleneck on the road to achieving true system-on-chip RF devices. This is the topic of Chap. 4.

Reflector type antennas were among the first structures to be expanded to millimeter regions, providing high gains and narrow beamwidths, and were well suited for radio astronomy, satellite communications, and high-speed microwave links. Chap. 5 discusses reflector structures and lens antennas and the numerous interesting configurations that have surfaced over the years, such as Fresnel zone plate lenses and millimeter-wave reflectarrays.

Antennas do not operate in isolation and the development of the electronics industry is a critical aspect of millimeter-wave systems. Following the discussions in the introductory sections of this text, Chap. 6 investigates the advances made in integrated circuit technology such as mixers, detectors, power amplifiers and oscillators. Passive circuitry such as switches and power combiners is also discussed.

To conclude this text, Chap. 7 provides detailed insight on applications that are made possible by millimeter-wave systems through several examples of recent developments in automotive radar, communication systems and defense applications. This chapter can therefore be approached with a firm idea of the progress that enabling technologies have made in recent years and the influence that these components have on millimeter-wave systems.

References

1. R. Dybdal, Millimeter wave antenna technology. IEEE J. Sel. Areas Commun. **1**(4), 633–644 (1983)
2. J.C. Wiltse, History of millimeter and submillimeter waves. IEEE Trans. Microw. Theory Tech. **32**(9), 1118–1127 (1984)
3. W. Gordy, Millimeter and submillimeter waves in physics, in *Proceedings of the Symposium on Millimeter Waves* (1959), pp. 1–22
4. A.K. Sen, Sir J.C. Bose and radio science. IEEE MTT-S Microw. Symp. Digest **2**, 557–560 (1997)
5. E.L. Ginzton, The $100 idea. IEEE Trans. Electron Devices **23**(7), 30–39 (1976)
6. R.H. Varian, S.F. Varian, A high frequency amplifier and oscillator. J. Appl. Phys. **10**(2), 321–327 (1939)
7. K. Campbell, SA radar sector has marked a major milestone in its history, *Creamer Media's Engineering News*, (2015). [Online]. Available: http://www.engineeringnews.co.za/print-version/sa-radar-sector-has-marked-a-major-milestone-in-its-history-2015-03-06. Accessed 27 July 2015
8. J.T. Randall, H.A. Boot, Historical Notes on the Cavity Magnetron, *IEEE Trans. Electron Devices*, **ED-23**(7), 724–729 (1976)
9. M. Skolnik, *"Introduction and Overview of Radar"*, in *Radar Handbook, New York City* (McGraw-Hill, New York, 2008)
10. R. Kompfner, The invention of traveling wave tubes. IEEE Trans. Electron Devices **23**(7), 730–738 (1976)
11. J.R. Pierce, History of the microwave-tube art. Proc. IRE **50**(5), 978–984 (1962)
12. G.C. Southworth, Principles and applications of waveguide transmission. Bell Syst. Tech. J. **29**(3), 295–342 (1950)
13. S. Schelkunoff, Conversion of Maxwell's equations into generalized telegraphist's equations. Bell Syst. Tech. J. **34**(5), 995–1043 (1955)
14. R. Beringer, The absorption of one-half centimeter electromagnetic waves in oxygen. Phys. Rev. **70**(1), 53 (1946)
15. A. Smith, W. Gordy, J. Simmons, W. Smith, Microwave spectroscopy in the region of three to five millimeters. Phys. Rev. **75**(2), 260 (1949)

16. W. Warters, WT4 millimeter waveguide system: Introduction. Bell Syst. Tech. J. **56**(10), 1825–1827 (1977)
17. F.B. Dyer, E.K. Reedy, Millimeter RADAR at Georgia Tech, in *S-MTT International Microwave Symposium Digest* (1974), p. 152
18. N.M. Kroll, M. Bernstein, Magnetron research at Columbia radiation laboratory. Trans. IRE Prof. Gr. Microw. Theory Tech. **2**(3), 33–37 (1954)
19. G. Collins, *Microwave Magnetrons*, vol. 6 (McGraw-Hill, Chicago, Illinois, 1948)
20. B.E.O. Willoughby, A. Member, E.M. Williams, Attenuation curves for 2:1 rectangular, square and circular waveguides, *J. Inst. Electr. Eng. - Part IIIA Radiolocation*, **93**(4), 723–724 (1946)
21. S.E. Miller, A.C. Beck, Low-loss waveguide transmission, *Proc. IRE* **41**(3), 348–358 (1953)
22. H. Unger, Helix waveguide theory and application. Bell Syst. Tech. J. **37**(6), 1599–1647 (1958)
23. S. Schlesinger, D. King, Dielectric image lines, *IEEE Trans. Microw. Theory Tech.* **MTT-6** (3), 291–299 (1958)
24. D.D. King, Properties of dielectric image lines, *IRE Trans. Microw. Theory Tech.*, **MTT-3**(2), 75–81 (1955)
25. D.D. King, Circuit components in dielectric image lines, *IRE Trans. Microw. Theory Tech.* **MTT-3**(6), 35–39 (1955)
26. J.C. Wiltse, Some characteristics of dielectric image lines at millimeter wavelengths, *IRE Trans. Microw. Theory Tech.* **7**(1) (1959)
27. F. Sobel, F. Wentworth, J.C. Wiltse, Quasi-optical surface waveguide and other components for the 100- to 100-gc region, *IRE Trans. Microw. Theory Tech.*, **MTT-9**(6), 512–518 (1960)
28. F. Tischer, Properties of the H-guide at microwave and millimetre-wave regions, *Proc. IEE Part B Electron. Commun. Eng.* **106**(13), 47–53 (1959)
29. F.J. Tischer, M. Cohn, Attenuation of the HE11 mode in the H-guide. IRE Trans. Microw. Theory Tech. **7**(4), 478–480 (1956)
30. S. Chatterjee, R. Chatterjee, Dielectric loaded waveguides—A rreview of theoretical solutions. Radio Electron. Eng. **30**(4), 195–205 (1965)
31. W.M. Sharpless, Wafer-type millimeter wave rectifiers. Bell Syst. Tech. J. **35**(6), 1385–1402 (1956)
32. M. Cohn, F. Wentworth, J.C. Wiltse, High-sensitivity 100- to 300-Gc radiometers. Proc. IEEE **51**(9), 1227–1232 (1963)
33. W. Johnson, T. Mori, F. Shimabukuro, Design, development, and initial measurements of a 1.4-mm radiometric system. IEEE Trans. Antennas Propag. **18**(4), 512–514 (1970)
34. S. Adachi, R. Rudduck, C. Walter, A general analysis of nonplaner, two-dimensional luneberg lenses. IRE Trans. Antennas Propag. **9**(3), 353–357 (1961)
35. P. Hannan, Microwave antennas derived from the cassegrain telescope, *IRE Trans. Antennas Propag.* **9**(2) (1961)
36. M. Gunn, Microwave conductivity of germanium. Proc. IEEE **52**(7), 851 (1964)
37. B.W. Knight, G.A. Peterson, Theory of the gunn effect. Phys. Rev. **155**(2), 393–404 (1967)
38. J. Copeland, CW operation of LSA oscillator diodes-44 to 88 GHz. Bell Syst. Tech. J. **46**(1), 284–287 (1967)
39. D. BenDaniel, C. Duke, Space-charge effects on electron tunneling. Phys. Rev. **152**(2), 683 (1966)
40. Y. Shiau, Dielectric rod antennas for millimeter-wave integrated circuits (short papers). IEEE Trans. Microw. Theory Tech. **24**(11), 869–872 (1976)
41. D.N. Black, J.C. Wiltse, Millimeter-wave characteristics of phase-correcting fresnel zone plates. IEEE Trans. Microw. Theory Tech. **35**(12), 1122–1129 (1987)
42. D.M. Pozar, *"Electromagnetic Theory", in Microwave Engineering*, 4th edn. (John Wiley & Sons Inc, Hoboken, New Jersey, 2012)
43. N. Guo, R.C. Qiu, S.S. Mo, K. Takahashi, 60-GHz Millimeter-wave radio: Principle, technology, and new results, *Eurasip J. Wirel. Commun. Netw.* **2007** (2007)

44. J.E. Freehafer, D.E. Kerr, *"Elements of the problem", in propagation of short radio waves, Chicago* (McGraw-Hill, Illinois, 1990)
45. Z. Pi, F. Khan, An introduction to millimeter-wave mobile broadband systems. IEEE Commun. Mag. **49**(6), 101–107 (2011)
46. B. Mahafza, *"Radar Systems—An Overview", in Radar Signal Analysis and Processing Using MATLAB, New York City* (CRC Press, New York, 2009)
47. J.S. Ojo, M.O. Ajewole, S.K. Sarkar, Rain rate and rain attenuation prediction for satellite communication in Ku and Ka bands over Nigeria. Prog. Electromagn. Res. B **5**, 207–223 (2008)
48. R.C. Daniels, R.W. Heath, 60 GHz Wireless communications: Emerging requirements and design recommendations. IEEE Veh. Technol. Mag. **2**(3), 41–50 (2007)
49. P. Smulders, A. Wagemans, Wideband indoor radio propagation measurements at 60 GHz. Electron. Lett. **28**(13), 1270–1272 (1992)
50. T. Manabe, K. Sato, H. Masuzawa, K. Taira, T. Ihara, Y. Kasashima, K. Yamaki, Polarization dependence of multipath propagation and high-speed transmission characteristics of indoor millimeter-wave channel at 60 GHz. IEEE Trans. Veh. Technol. **44**(2), 268–274 (1995)
51. R.K. Crane, Propagation phenomena affecting satellite communication systems operating in the centimeter and millimeter wavelength bands, *Proc. IEEE* **59**(2) (1971)
52. Y.P. Zhang, P.C. Ching, Y. Hwang, Theory of millimeter-wave propagation in railway trains, in *Vehicular Technology Conference, 1998. VTC 98. 48th IEEE (Volume:1)* (1998), pp. 530–533
53. W. Keusgen, R.J. Weiler, M. Peter, M. Wisotzki, Propagation measurements and simulations for millimeter-wave mobile access in a busy urban environment, *9th Int. Conf. Infrared, Millimeter, Terahertz Waves, IRMMW-THz 2014*, (14–19 Sept, Tucson, USA, 2014)
54. K. Sato, H. Kozima, H. Masuzawa, T. Manabe, T. Ihara, Y. Kasashima, K. Yamaki, Measurements of reflection characteristics and refractive indices of interior construction materials in millimeter-wave bands, *1995 IEEE 45th Veh. Technol. Conf. Countdown to Wirel. Twenty-First Century*, **1** (1995)

Chapter 2
Leaky-Wave Antennas

Millimeter waveguides, such as coplanar lines, microstrip lines and dielectric guides, are open structures where energy leakage is bound to occur. This leakage can occur, for example, when there are discontinuities in the guiding structure, or perhaps when the guide is excited in an inappropriate mode. In terms of antenna design, this energy leakage may be advantageous and exploited. This is accomplished through intentionally creating discontinuities in the guiding structure in such a manner that the radiation produced is controlled by the antenna designer. These antennas are very suitable for integrated designs, seeing that they are compatible with the waveguides from which they are derived, essentially limiting the unwanted and imperfect transitions between guiding media. This property also eliminates the requirement for complex and lossy feed networks present in other types of planar structures.

Mathematically, a leaky wave is treated as a guided complex wave and the resulting radiation pattern is expressed in terms of the complex propagation constant. As a subset of traveling wave antennas, leaky-wave antennas are further divisible into two categories, namely one-dimensional and two-dimensional variants. These will be further reduced into subclasses and introduced as this chapter progresses. Similar to other traveling wave antennas, leaky-wave antennas radiate primarily in the endfire direction, but they are very suitable for frequency scanning and as such are often implemented for this purpose. Surface wave antennas and slot arrays are also part of the traveling wave family, and while they share some defining features, their performance expectations and design methodologies differ.

2.1 General Principles of Leaky Waves

The properties of leaky waves were originally derived in the pioneering work of Oliner and Tamir in the late 1950s and early 1960s [1–3]. This was followed by an extensive development of leaky-wave theory and application to antennas. However,

© Springer International Publishing Switzerland 2016
J. du Preez and S. Sinha, *Millimeter-Wave Antennas: Configurations and Applications*, Signals and Communication Technology,
DOI 10.1007/978-3-319-35068-4_2

interest in the behavior and application of these antennas at millimeter-wavelengths only began several decades later.

As mentioned earlier, a leaky wave arises from a guiding structure with some or other continuous or periodic discontinuity that facilitates energy leakage into the surrounding area. A simple example is a slotted waveguide antenna, where the waveguide is perturbed with periodic slots in the structure at a certain position. The energy leakage mechanism results in the waveguide having a propagation wave number that is a complex quantity. The phase constant is given by β and the attenuation constant is indicated by α. The attenuation constant varies in size depending on the leakage per unit length along the waveguide; a larger value for α means that a larger amount of energy leaks from the structure [4].

As seen from Fig. 2.1, the complex propagation parameters are dependent on the geometry of the leaky structure. Control over the beam shape and sidelobe levels may be obtained by using an appropriate aperture taper. This is accomplished in practice by slowly varying α along the length of the guide in a specific fashion, and simultaneously holding the phase constant at a fixed value, effectively adjusting the amplitude of the aperture distribution.

The entire waveguide in Fig. 2.1 is regarded as the effective aperture of the antenna, unless the energy leakage is so severe that the signal power fades away completely before reaching the end of the slotted section. A large value for α (which indicates a large leakage rate) in turn leads to a smaller effective aperture, and from antenna theory it is known that the smaller aperture would lead to an increase in beamwidth [5]. On the contrary, lower α values result in a much longer effective aperture and thus the antenna would radiate a narrow main beam, if the physical size of the structure permits it.

It should be noted that if the aperture is fixed beforehand and the leakage rate α is relatively small, the antenna pattern (particularly the beamwidth) is primarily influenced by the aperture itself, rather than the leakage rate. However, the radiation efficiency in such a case is strongly affected by the leakage rate. A reasonable design goal is to have approximately 90 % of the power in the waveguide radiated from the structure when the wave has propagated through the entire structure. In most practical cases, a matched load will be connected at the end of the waveguide in order to absorb the remainder of the signal power.

Fig. 2.1 Slotted rectangular waveguide leaky-wave antenna

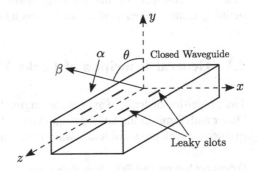

Rectangular waveguides support either transverse electric (TE) or transverse magnetic (TM) modes of wave propagation, and for both cases the phase constant of the wave is frequency dependent [6]. As the frequency (and so also the phase constant β) changes, so does the direction in which the principal beam is pointing. Therefore, the antenna beam can be scanned by altering the excitation frequency.

2.2 Extension into Millimeter-Wavelengths

Leaky-wave antennas have been extensively studied in the last three to four decades, and their main attraction is high directivity, wide bandwidths, and their ability to scan with frequency. Earlier, leaky-wave antennas were based almost exclusively on closed waveguides and the leakage was introduced by cutting holes or slots into the guiding structure. Waveguide losses become exceedingly high at millimeter-wavelengths. As a result, millimeter waveguides are typically open structures, in an attempt to lower the attenuation constant that results from dielectric or metallic losses [7]. Examples of open waveguides are microstrip lines, groove guides, non-radiative dielectric guide and many other configurations of dielectric guides. Generally, the dominant modes that propagate in these types of waveguides are purely bound, which means that physical defects in the structure will not cause them to radiate. Instead, other techniques such as asymmetry or similar geometric alterations are often used.

Simple open waveguides support the propagation of slow waves, which do not radiate outward from the guide, and the periodic structure is chosen such that only the $n = -1$ space harmonic is able to radiate power. On the other hand, some millimeter waveguides can be closed structures, such as metal guides and finlines. The tapered antennas shown in Fig. 2.4, particularly, are examples of antennas that are easily integrated with the waveguides from which they are derived [8, 9].

With a basic theoretical background on leaky-wave modes established, we will now move on to covering influential designs and concepts that have surfaced in recent years—in no particular order.

2.3 Leaky-Wave Antenna Classification

Depending on the geometry, principle of operation and the nature of structural perturbations, leaky-wave antennas can be divided into several categories [10]. As mentioned in the introduction, perhaps the most basic distinction is between one- and two-dimensional leaky-wave antennas. Furthermore, these can be identified as periodic, uniform or quasi-uniform structures. These configurations will be discussed separately in the succeeding sections, but it should be noted that some of the concepts relating to either the theory of operation or design principles overlap, sometimes with minor differences.

2.3.1   One-Dimensional Uniform Leaky-Wave Antennas

Theory

A one-dimensional structure is one that supports a wave traveling in a single, fixed direction. One such structure is a rectangular waveguide with a long uniform slit in one of its side walls, like the one shown in Fig. 2.2. The geometry of this guide can be considered uniform, seeing that it does not vary in the longitudinal direction (along the z-axis). The slot in Fig. 2.2 is modeled with a surrounding infinite ground plane that acts as a back-baffle, and its shape may be tapered if a specific beam shape is desired. For a narrow slot, the structure depicted in Fig. 2.2 becomes equivalent to a magnetic line current in the direction of the z-axis [4].

The radiation produced by this structure is limited to the $z > 0$ region, and the resulting beam is in the shape of a cone along the z-axis. As the angle approaches 90° (indicated by θ in Fig. 2.1)—in other words, as one approaches the broadside direction—the antenna pattern takes on a narrow donut-type shape.

It is difficult to achieve a broadside beam if the antenna is fed from only one end, since this equates to operating the waveguide at its cutoff. By feeding the waveguide with a frequency that is marginally higher than its cutoff, it may be possible to achieve a broadside beam. This may be accomplished by feeding it from both ends, or by placing the source in the middle of the guiding structure. From the examples provided here, it is clear that these antennas are axially significantly longer than laterally.

An extremely narrow pencil beam in both the azimuth and elevation planes can be obtained by placing the one-dimensional leaky-wave elements in a linear array [11]. The direction in which the beam points is then controlled by individual beam angles and the phase delta between elements. It is therefore possible to control the beam angle in the elevation plane by altering the excitation frequency and in the azimuth plane by altering the phase of the exciter.

The radiation pattern is obtained by taking the Fourier transform of the aperture distribution. In the case where the geometry is kept consistent along the length of the antenna, the field lines comprise a traveling wave with constant values for α and β. This then leads to an exponentially decaying amplitude distribution along the length of the guide. If the length of the antenna is modeled as being infinite, the resulting radiation pattern can be computed with acceptable accuracy as

Fig. 2.2 Rectangular waveguide with a longitudinal slot on its side wall

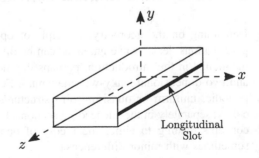

Longitudinal Slot

$$R(\vartheta) \approx \frac{\cos^2 \vartheta}{\left(\frac{\alpha}{k_0}\right)^2 + \left(\frac{\beta}{k_0} - \sin \vartheta\right)^2}, \tag{2.1}$$

where k_0 is the free-space wavenumber. Fundamentally, this pattern does not contain any sidelobes, but as the length is decreased the expression for $R(\vartheta)$ becomes more involved and sidelobes begin to appear. The sidelobes present in a leaky-wave antenna are often significant, and the tapering approach to its design is often the de facto route that designers follow in practice.

Design Principles

Once the values for α and β are known, the major parameters of the antenna—beamwidth, radiation efficiency, and scan angle—can be determined rapidly. The scan angle is given by

$$\sin \theta_{p,max} \approx \frac{\beta}{k_0}, \tag{2.2}$$

where, as the subscript indicates, $\theta_{p,max}$ represents the maximum achievable scan angle. For a waveguide with axial length L, the beamwidth can then be found as

$$\Delta\theta \approx \frac{1}{(L/\lambda_0) \cos \theta_{p,max}}. \tag{2.3}$$

The unity factor in the numerator of Eq. (2.3) changes depending on the particular amplitude distribution. For example, for an aperture distribution that remains consistent over the length of the antenna, the numerator is replaced by 0.88, while a tapered distribution could have a numerator greater than 1.25 [4]. The physical length of the antenna L is chosen so that approximately 90 % of the power is radiated. The remainder is then absorbed by a matched load, and L is usually specified for a set value of α. Therefore, we can find L as

$$\frac{L}{\lambda_0} \approx \frac{0.18}{\alpha/k_0}, \tag{2.4}$$

where L and α are chosen independently of one another, the percentage of radiated power can differ significantly from the desired value of 90 %. The ratio of power that resides in the leaky mode at $z = L$ to the input power can then be written as

$$\frac{P(z = L)}{P(z = 0)} = \exp(-2\alpha L) = \exp(-4\pi(\alpha/k_0)(L/\lambda_0)), \tag{2.5}$$

where $P(z)$ indicates the power in the waveguide at a distance z along the length of the structure. The radiated power changes slightly when the beam is scanned with frequency, given that α changes with frequency. However, the percentage of radiated power (indicated simply by P%) can be obtained easily by rewriting Eq. (2.5), assuming an exponential aperture distribution. The result is

$$P\% = 100\{1 - \exp[-4\pi(\alpha/k_0)(L/\lambda_0)]\}. \tag{2.6}$$

The aperture distribution will inevitably be changed in order to control sidelobes, but nonetheless, Eq. (2.6) remains a reasonable approximation.

In terms of scan angle, uniform leaky-wave antennas can take on two forms. While the principles remain similar, the scan angle behavior of air-filled and dielectric-filled waveguides differ somewhat. Air-filled structures commonly encountered are rectangular waveguides and groove guides, both of which support dominant modes that are fast. On the other hand, partially dielectric-filled structures often used are the non-radiative dielectric guide and dielectric-loaded rectangular guide. To operate these as leaky-wave antennas, they should be excited with a fast wave ($\beta < k_0$).

Research Review

Considering uniform leaky-wave antennas, a lot of effort has been expended towards non-radiative dielectric guides, often referred to simply as NRD guides. Early work by Oliner et al. contained detailed theories and experimental results on rectangular NRD guides, although several other variations of leaky-wave NRD guides have been proposed for millimeter-wavelengths ever since. The first proposal on leaky-wave antennas derived from NRD guides appeared in 1981 and was authored by Yoneyama et al. [12].

The concept of asymmetrically perturbed planar leaky-wave antennas has been studied extensively by Gómez-Tornero et al. [7]. This work covered several important questions relating to planar leaky-wave antennas, and the frequencies of interest were around 40–65 GHz. For instance, modifying the leakage factor of the leaky-wave mode while keeping the phase constant stationary was investigated for a wide array of slot and strip designs.

Each of these designs used a different tapering topology and was specifically designed to be mechanically flexible and thus relatively simple to fabricate. One example of a tapered planar guide is shown in Fig. 2.3, where the width, amplitude

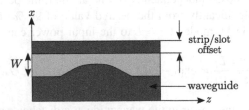

Fig. 2.3 Tapered slot leaky-wave antenna. The offset, shape, and maximum width of the taper are controlled to alter radiation characteristics

distribution, and the offset of the slot (or microstrip in some cases) affect the properties of the leaky-wave modes propagating in the guide.

Through the use of a hybrid-planar topology, flexibility in manufacturing became a substantial benefit when compared to conventional rectangular waveguide technology. This is especially important at millimeter-wavelengths since manufacturing tolerances play a key role in the performance of the antenna. The authors came to the conclusion that the slotted leaky-wave antenna proved to be better suited for offset tapering, while the stripline leaky-wave antennas were in turn better suited for asymmetric tapering around a zero offset.

The group published its analysis and design methods for the hybrid waveguide-planar technology later that year. It provided full-wave integral equations and design guidelines for this concept [13]. The derivations were based on the leaky-wave modes that propagate in laterally shielded, stub-loaded dielectric guides that are rectangular in shape. The perturbations in the guiding structures were considered as either rectangular strips or slots. The guidelines presented in this report allowed other designers to obtain leaky-wave dispersion curves easily. A demonstration of this was provided by developing a novel planar leaky-wave antenna. The configuration presented had the attractive property of being able to alter the stopband region and the leakage rates independently, because of relying on the principle of asymmetric radiation.

Gómez-Tornero et al. [14] continued work with one-dimensional leaky-wave antennas, reporting on a parallel-plate waveguide loaded on one end with a partially radiating surface and on the other end with a high impedance surface. This antenna is designed for operation at 15 GHz. By altering the dimensions of the dipoles etched on the two surfaces, independent control over the leakage rate and the leaky mode phase constant is possible. Using an air-filled waveguide also negates the dielectric loss, leading to radiation efficiencies of about 90 %.

Furthermore, the group also investigated conformal tapered microstrip leaky-wave antennas. These were based on using half-width microstrip sections at 15 GHz [15].

2.3.2   One-Dimensional Periodic Leaky-Wave Antenna

Theory

A one-dimensional periodic antenna comprises a uniform structure that supports the propagation of a non-radiating wave (i.e., $\beta > k_0$) and is modified periodically in the z-direction. This non-radiating wave is also referred to as a fast wave, since its dominant mode is fast relative to free-space velocity. In a periodic antenna, however, the dominant mode is a slow wave and therefore does not radiate by itself: the structure needs some kind of periodic modulation in order to produce the radiation.

An example is shown in Fig. 2.4, where slots are periodically inserted on a planar structure and the slot length is controlled through a taper function. This

Fig. 2.4 Tapered slot
leaky-wave antennas, **a** using
an exponential taper and
b using a linear taper

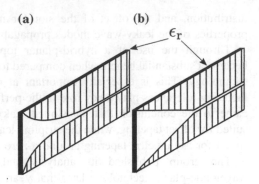

configuration differs from the tapered slot antenna shown in Fig. 2.3, in that the slots are periodically inserted on the structure, although both are essentially planar configurations. The tapering is implemented differently and the effects on the antenna characteristics should become clear as the discussion progresses.

A primary advantage of periodic leaky-wave antennas is that the direction in which the beam points could be either forwards or backwards, depending on the phase constant used in the excitation.

Although two planar structures are shown in Fig. 2.4, the concept applies to many other periodic configurations. For example, a dielectric waveguide can be periodically perturbed by small holes or slots and achieve similar radiation characteristics. The excitation of the waveguide is in the fundamental mode, and in order to prevent higher order mode propagation, the width is chosen to be relatively small.

The slots in Fig. 2.4 could also be replaced with a metal grating (i.e., periodic metal strips on the dielectric) and similarly a dielectric grating. In the latter case, the metal strips would instead be grooves in the dielectric surface, and the diffraction effect at this grating would then transform the excitation mode into a leaky wave [16].

Since the radiation mechanisms of uniform and periodic leaky-wave antennas result from different physical processes, these antennas differ in their achievable scan ranges. Consider the dielectric waveguide in Fig. 2.5.

The guide dimensions are chosen such that the dominant mode is the only mode above the cutoff frequency, and since $\beta > k_0$ (equivalently, $\beta_n/k_0 < 1$) for this particular mode, it is thus purely bound. After the dimensions have been set, the metal strips in Fig. 2.5 are added periodically, and it is this periodicity that in turn

Fig. 2.5 Dielectric rod, periodically loaded with metallic strips

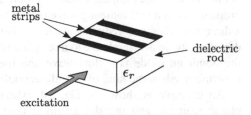

creates an infinite number of space harmonics. These harmonics are characterized by a unique phase constant β_n and subsequent phase constants are then related by

$$\beta_n d = \beta_0 d + 2n\pi, \tag{2.7}$$

where d is the strip spacing and β_0 indicates the original phase constant of the dominant mode in a uniform guide [4]. The phase constant β_n can clearly assume a significant number of values, but if at least one of the space harmonics becomes fast, the whole mode will become leaky. Given that $\beta_n/k_0 < 1$ and $\beta_0/k_0 > 1$, the relationship in Eq. (2.7) can then be rewritten as

$$\frac{\beta_n}{k_0} = \frac{\beta_0}{k_0} + \frac{2n\pi}{k_0 d} \tag{2.8}$$

and since $k_0 = 2\pi/\lambda_0$, it is clear that it is indeed possible for $|\beta_n/k_0|$ to be less than unity, provided that λ_0/d is suitably selected and the harmonic number n is negative. For an antenna that is designed to have only one radiated beam, one can go ahead and choose $n = -1$. Following this approach, at low frequencies there will now be fast harmonics and the antenna will not radiate any beams. As the frequency increases and eventually reaches a critical point where the $n = -1$ harmonic becomes fast, a beam originates from backward endfire. Further increasing the frequency will result in the beam scanning from backward endfire, through broadside, up to forward endfire.

A periodic leaky-wave antenna therefore radiates as a result of the $n = -1$ space harmonic, which is the fundamental difference between uniform and periodic structures.

Design Principles

Tapering the antenna aperture according to the appropriate amplitude distribution leads to exceptional radiation characteristics and ultra-low sidelobes on paper. To obtain these properties in practice, careful design of the feeding method is required. When the antenna is derived from perturbing a closed waveguide, the taper will cause the aperture to radiate minimally at both ends. The discontinuity between the antenna and the closed feed waveguide is thus negligible, and considerations relating to the feed are nearly nonexistent.

This is not the case when the feed is an open waveguide. Surface waves are excited from a tapered transition out of a closed waveguide, and the transition creates spurious radiation. Wherever this type of transition is found in the feed network, the uncontrolled radiation may adversely affect the radiation pattern to the point where the initial design is invalidated. The influence that spurious radiation has on the radiation pattern is not an issue with most leaky-wave antennas, but given the effects on the antenna performance, it should not be overlooked in the design.

In order to ensure that the TE_{10} mode is a slow wave over the desired bandwidth, a lower limit on the permittivity of the substrate is given by

$$\epsilon_r > 1 + \left(\frac{\pi}{k_0 a}\right)^2, \tag{2.9}$$

where a indicates the greater of the two waveguide dimensions. Although there are many derivations of leaky-wave antennas based on rectangular waveguides, many properties are common and can be understood by considering this structure. The fundamental mode that exists in the waveguide is non-radiating, but the periodicity in the structure means that the modal field is in the form of a Floquet-wave expansion,

$$E(x, y, z) = \sum_{n=-\infty}^{\infty} A_n(x, y) \exp(-jk_{zn}z), \tag{2.10}$$

where

$$k_{zn} = k_{z0} + \frac{2n\pi}{p} \tag{2.11}$$

is the specific wave number for the nth Floquet mode—also often referred to as a space harmonic—recall the earlier discussion—and p indicates the perturbation period [10]. As mentioned earlier, leakage will occur if one of the space harmonics is a fast wave. This is usually the $n = -1$ harmonic, and the result is the requirement that $-k_0 < \beta_{-1} < k_0$. Altering the periodicity p in the appropriate manner can scan the main beam over to the desired angle and, typically, it can be scanned from backward endfire to forward endfire. Alternatively, beam scanning is achieved by changing the excitation frequency. Having a single beam over the scan range is an important design consideration. This occurs when the $n = -2$ harmonic remains as a slow backward wave ($\beta_{-2} < -k_0$), while the fundamental mode is kept as a slow forward wave ($\beta_0 > k_0$). At the highest frequency in the range of scan angles (that is, in the forward endfire direction), it is then required that

$$\epsilon_r > 9 + \left(\frac{p}{a}\right)^2, \tag{2.12}$$

which follows from $p/\lambda_0 < 0.5$. One thing to notice is that for uniform periodic leaky-wave antennas, a dielectric constant of greater than 9 is always required if a single beam is desired for the application. If the guiding structure does not support a fundamental mode, the constraint outlined in Eq. (2.12) changes. One example is a microstrip transmission line, which can support only a quasi-transverse electromagnetic (TEM) mode [6].

Research Review

The earliest forms of periodic one-dimensional leaky-wave antennas were in the form of dielectric rods (one such antenna was illustrated in Fig. 2.5). Most of these results were due to the pioneering work of Klohn et al. [11], and successful millimeter-wave designs were demonstrated as early as 1978. Particular bands of interest were in the 55–100 GHz region, specifically around 60, 70, and 94 GHz, millimeter bands that are widely used today. Though discussed extensively in the introductory chapter of this text, it is worth mentioning once again that these developments were highly dependent on the advent of stable, high-power millimeter-wave sources, many of which were based on the IMPATT diode. The work of Klohn et al. also clearly demonstrated the manufacturing challenges associated with millimeter-wave designs. For example, it was found that a 0.1 mm change in the perturbation period led to a shift in scanning angle of approximately 8°. Nonetheless, these antennas were a crucial part of the development of millimeter-wave integrated circuits.

Investigation into the endfire tapered slot antennas shown in Fig. 2.4 was first reported by the Yngvesson group [8], a few years after the publication of the research done by Klohn's group. With applications such as remote sensing, radio astronomy, and satellite communications, there was an increasingly urgent requirement for multi-beam antennas and integrated systems. The Vivaldi antenna presented by Gibson in 1979 is an early example of a planar endfire antenna that exhibits acceptable gain and sidelobe levels. The Vivaldi antenna uses an exponential taper to control the amplitude distribution of the individual slots, and the configuration shown in Fig. 2.4a is thus a Vivaldi antenna. One substantial advantage offered by tapered slot antennas is their ability to produce symmetric E-plane and H-plane beams over large bandwidths, even with the planar implementation discussed here.

A group at the Valencia Polytechnic University has investigated the possibilities of constructing leaky-wave antennas that are based on the Goubau line [17].

High losses in the dielectric and ground plane of microstrip circuits offer a substantial challenge at millimeter-wavelengths, and other types of transmission lines are often required to solve this problem. At submillimeter wavelengths, single-wire waveguides have been demonstrated and validated for reasonably low-loss transmission; however, these lines are far too complex in structure to facilitate their practical use in array configurations. On the other hand, the Goubau line is an effective alternative to the single-wire waveguide. In fact, the variation discussed here is simply a planar implementation of the thin-wire waveguide on a thin substrate. This is shown in Fig. 2.6.

Similar to the traditional microstrip transmission line, the Goubau line does not use a ground plane (as illustrated in Fig. 2.6) and as a result, presents a lower attenuation as opposed to that of the microstrip line. The leaky-wave configuration that is based on the Goubau line is shown in Fig. 2.7.

The fundamental mode that propagates along the Goubau line is bounded. It can therefore be used as a leaky-wave antenna by adding dipole sources along the

Fig. 2.6 The Goubau transmission line in its planar form

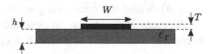

Fig. 2.7 Transverse dipole array with a Goubau transmission line. Note once again the absence of a ground plane

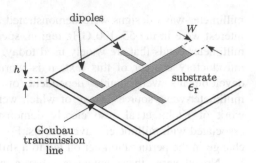

transmission line. The transverse planar dipole sources are periodic perturbations along the line and they give rise to an infinite number of space harmonics. If a single one of these harmonics is fast, it will cause the structure to radiate. The separation distance of the dipoles is chosen such that the main beam is not radiating in the broadside direction, since this significantly decreases the gain of the antenna. To ensure that the structure radiates a single main beam, the mode $n = -1$ is selected. The final configuration consists of a 16-element array (8×2) with a cosine distribution.

It was later discovered that the absence of a ground plane resulted in two main lobes, as opposed to the intended single-beam implementation. To realize the single-beam version, a metallic plate was added to the structure at an optimized distance from the antenna. This optimization was required so that the power radiated towards the metallic plate would be completely reflected as well as added in-phase to the generated antenna beam. Adding the two beams in-phase resulted in a maximum obtainable directivity. Furthermore, reduced losses could be achieved if the metallic plane and the antenna were separated by an air gap. Measurements were taken around 40 GHz, particularly in the frequency band of 37–41 GHz. The obtainable radiation efficiency was found to be 71 % (70 % without the reflecting metallic plate), and S_{11} remained below -10 dB over the specified operating bandwidth.

A later modification to the Goubau line approach presented by Sánchez-Escuderos et al. [17] involved the design of arrays that were circularly polarized. This was accomplished by printing crossed dipoles on the substrate with an electric contact between the elements. The measured H-plane radiation pattern obtained remained properly circularly polarized with and without the metallic reflector plane discussed in the earlier work.

In a subsequent publication, Sánchez-Escuderos et al. [18] reported detailed measurements on the circularly polarized variation of the Goubau line leaky-wave antenna. The measured results obtained from the prototype antenna were promising, with a 15 % impedance bandwidth (for $S_{11} < -10$ dB) and a 3-dB axial ratio bandwidth of 7.6 %. Over the specified bandwidth, the radiation efficiency was over 90 % and the measured gain of the antenna was found to be 15.6 dBi.

A recent investigation on periodic leaky-wave antennas by Nechaev et al. [19] compared two types of center-fed configurations for millimeter-wave operation.

The first of these consisted of a periodically grated grounded dielectric waveguide. The grating comprised two arrays of parallel metallic strips, and the design frequency was centered at 81 GHz.

2.3.3   Two-Dimensional Leaky-Wave Antenna

Theory

A leaky wave that originates from a two-dimensional guiding structure propagates radially from the feed point. This configuration provides an easy method of obtaining a directive beam at broadside, requiring only a simple source. The general form of such a structure is a partially reflecting surface above a ground substrate. This is illustrated in Fig. 2.8.

The excitation source shown in Fig. 2.8 is a simple horizontal dipole placed within the substrate, at a set distance above the ground plane. The antenna pattern, however, depends on the structure and not on the excitation. The substrate/superstrate structure could also be extended to include multiple layers of the dielectric material, with the advantage of narrowing the beamwidth [10]. Increasing the permittivity of the superstrate layer results in a narrower beamwidth.

Another example of a partially reflecting surface is shown in Fig. 2.9.

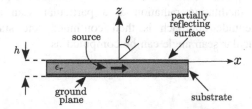

Fig. 2.8 Substrate/superstrate structure of a two-dimensional leaky-wave antenna

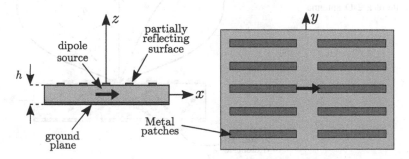

Fig. 2.9 Construction of a partially reflecting surface using metal patches

Larger rectangular microstrip elements will result in a narrower beamwidth. This structure is very suitable for photolithographic fabrication. A similar configuration to the one shown in Fig. 2.9 is to replace the patches with slots. Conversely, smaller slots will result in a narrower beamwidth, and it should be noted that the surface wave feeding the antenna is directed in the y-direction (rather than in the x-direction as with the microstrip array).

As the excitation frequency is increased, the pencil beam originating from the antenna will assume a conical shape, effectively losing gain in the broadside direction [20]. Generally, the guiding structure in a two-dimensional leaky-wave antenna is uniform (or at least quasi-uniform) and the excited wave is a fast wave.

Design Principles

Two popular methods of producing frequency scanning pencil beams from a leaky-wave antenna is using an array of linear or uniform antennas, or by using a two-dimensional structure. While either of these approaches will produce highly directive beams, it is often required to stack multiple layers of dielectrics or use layers with exceedingly high permittivity. In order to improve on this technique, the work reported by Zhao et al. [21] describes a two-dimensional structure that consists of a periodic array of patch elements on a grounded dielectric substrate. Such a structure was illustrated in Fig. 2.9, and the design of such an antenna will be discussed in greater detail here.

As mentioned previously in this chapter, the partially radiating surface causes the antenna to behave like a leaky parallel-plate waveguide operating in the $n = 1$ mode. The array of patch elements acts as a leaky conducting plate on top of the dielectric surface, facilitating radiation at a particular scan angle θ_p. If the parallel-plate waveguide approach is then continued, the substrate thickness required for a particular scan angle can be computed as

$$k_{z1}h = (k_1 \cos \theta_d)h = n\pi \tag{2.13}$$

Fig. 2.10 Propagation of a leaky wave inside the substrate of a 2-D antenna

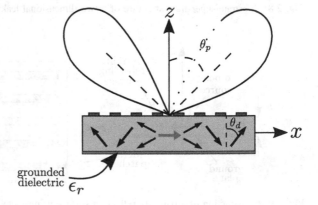

grounded dielectric ϵ_r

where n = 1, and θ_d indicates the angle at which the waves contained in the dielectric propagate. These are related to the radiation angle described by Snell's law [22] and this effect is detailed in Fig. 2.10.

From the figure, it should be clear that a broadside beam would correspond to $\theta_p = 0°$, and a main beam in the endfire direction corresponds to $\theta_p = 90°$.

The thickness of the substrate is then given by

$$\frac{h}{\lambda_0} = \frac{0.5}{\sqrt{\epsilon_r - \sin^2 \theta_p}}. \tag{2.14}$$

The spacing of patch elements should be small enough to result in a reflection coefficient of -1 from the radiating surface. Once the substrate thickness is found, the required permittivity can be computed.

A wide range of substrate permittivity values will result in a pencil beam and it is therefore a relatively arbitrary parameter. However, the permittivity heavily influences the scanning ability of the antenna. In a case where the substrate permittivity is chosen too low, it is very likely that a critical angle will exist at which the structure will produce secondary beams due to the excitation of higher order modes. This is highly undesirable, and in order for the beam to be able to scan at any angle between zero and endfire (i.e., $0 < \theta < 90°$), the n = 1 mode must scan over the whole range of interest before the next higher order mode appears at broadside.

For the principal higher order mode (n = 1) to produce endfire radiation, the substrate thickness can be found by entering the desired angle into Eq. (2.14), more specifically,

$$\frac{h}{\lambda_0} = \frac{0.5}{\sqrt{\epsilon_r - 1}}. \tag{2.15}$$

Furthermore, for the n = 2 mode to produce broadside radiation, the required substrate thickness can be found as

$$\frac{h}{\lambda_0} = \frac{1}{\sqrt{\epsilon_r}}. \tag{2.16}$$

Equating Eqs. (2.15) and (2.16), a value for the permittivity is found as $\epsilon_r = 4/3$, which is effectively the lower limit that will result in a single beam that is achievable at any given scan angle. The maximum achievable scan angle can then be determined by first finding the required substrate thickness when the n = 2 mode is excited, and then proceeding to evaluate Eq. (2.14) to find the scan angle. The maximum achievable scan angle (without exciting higher order modes) is then given by

$$\theta_p = \sin^{-1} \frac{\sqrt{3\epsilon_r}}{2}. \tag{2.17}$$

The last remaining design parameter is the spacing between patch elements. In order to prevent grating lobes existing at any given scan angle, the element spacing should be constrained. This value can be found by setting the $n = -1$ mode at backward endfire, while the main $n = 0$ mode is at forward endfire. The $n = 0$ mode will thus be at the forward endfire position when

$$\sqrt{k_1^2 - \left(\frac{\pi}{h}\right)^2} = k_0. \tag{2.18}$$

Furthermore, the $n = -1$ mode radiates in the backward endfire direction when

$$\sqrt{k_1^2 - \left(\frac{\pi}{h}\right)^2} - \frac{2\pi}{a} = -k_0, \tag{2.19}$$

where a indicates the element periodicity. Once again enforcing these two equations, a limit on the periodicity that will prevent grating lobes is established as $a/\lambda_0 < 0.5$. This relationship is commonly found in array design.

Research Review

The proposed results were obtained around 12 GHz, but nonetheless the metal patch array leaky-wave antenna proposed and investigated by Zhao et al. has inspired several other designers in the millimeter region to approach leaky-wave antennas in a different manner. This approach is essentially an alternative to dielectric layer leaky-wave antennas [23, 24]. The main advantage offered by the metal patch approach is in terms of directivity. For the dielectric layer antenna to achieve very high directivity, the substrate permittivity needs to be exceedingly high. On the other hand, it is far simpler to obtain high directivity with the patch method.

Slot arrays are an important class of millimeter-wave antennas, and the structure is based on cutting slots with different shapes and rotations into the walls of a waveguide [25]. This technique can be implemented with rectangular waveguides, or with substrate integrated waveguide technology.

One such slot array has been proposed by Zhao et al. [26]. Both the metal patch and slot array configurations were able to achieve a scan range of approximately 60°, if an air substrate was used. However, upon substituting air with a dielectric substrate, scanning to the horizon was made possible in both the E- and H-planes. For practical dimensions, it is simpler to generate narrower beamwidths by using a slot array, since there is effectively less leakage.

Originally investigated by Xu et al. [27], the concept of deriving leaky-wave antennas from substrate integrated waveguides is essentially a continuation of the earlier work. A substrate integrated waveguide is an extremely popular transmission

Fig. 2.11 Leaky-wave
antenna derived from a
periodically slotted substrate
integrated waveguide

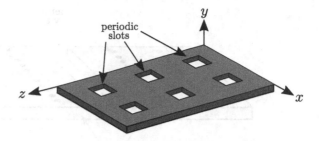

line at microwave frequencies and by constructing such an open periodic waveguide with large spacing between vias (or slots) can be used for leaky-wave antenna
design. A novel design and simulation approach was proposed, and the measured
return loss was lower than −15 dB over the 28–34 GHz band (a 6 GHz bandwidth).
An example of such a slotted array built on a substrate integrated waveguide is
shown in Fig. 2.11.

The work also illustrated the differences between the leaky-wave properties of
the TE_{20} and TE_{10} modes, concluding that the TE_{20} mode exhibits improved
radiation properties compared to the TE_{10} mode. Discussed earlier in this chapter,
the design highlighted the importance of substrate permittivity and slot periodicity
with regard to radiation efficiency, directivity, and frequency scanning capability.

Several other reports on leaky-wave antennas based on substrate integrated
waveguides have surfaced in recent years. A slot array that uses a plastic substrate,
fabricated using micromachining and injection molding techniques, has been
demonstrated by Fuh et al. [28]. The input flange is integrated directly into the
waveguide through a one-shot molding process, significantly increasing the consistency and robustness of the manufacturing process. The design of this antenna
explores many interesting manufacturing techniques. First, the waveguide is molded from a polymeric material through an injection molding process, whereafter a
metallic layer is deposited onto the waveguide through a sputtering and electro-
plating process. A self-alignment mechanism is introduced between the top metallic
plate and the plastic waveguide by using precision alignment pins. Finally, the feed
connectors are constructed from an integrated flange that is molded together with
the plastic waveguide. All of these features serve to reduce manufacturing costs and
this antenna provides a flexible architecture for radar designers.

Measurements of this antenna were taken in the W-band region (73–79 GHz).
To investigate the effects of surface roughness on the signal attenuation, a surface
profiler was used to scan the waveguide area. An RMS surface roughness of
1.0 ± 0.3 μm was found, which is higher than that of most commercially available
metallic waveguides, which generally exhibit a surface roughness of around 0.8 μm
on copper. The −10 dB bandwidth was measured as 9.3 GHz, with a 9.6 dB gain
and −13.5 dB sidelobe levels.

Another radar antenna based on substrate integrated technology was recently
proposed by Cheng et al. [29]. The system consists of a monopulse comparator, a
16-way power divider network and a 32 × 32 slot array, all integrated on a single

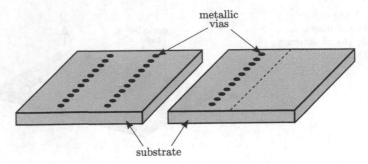

Fig. 2.12 Substrate integrated waveguide; the regular configuration is shown on the *left*, with the half-mode variant on the *right*

substrate. Monopulse antennas are particularly attractive for high-resolution tracking applications, and this antenna provides a highly directive beam and operates around 93–96 GHz. The system is etched on a single Rogers RT/Duroid 5880 substrate with a thickness of 0.508 mm and a dielectric constant of $\epsilon_r = 2.2$. The final assembly is 130×125 mm in size.

As expected from such a large array, the antenna produces a narrow beamwidth of 2°–3° in the E-plane and 3°–4° in the H-plane, when excited at the sum port. The minimum and maximum values of the measured gain vary between 21.29 dBi at 96 GHz and 25.75 dBi at 93.6 GHz. Sidelobe levels varied between 6 and 20 dB in the H-plane and 10–12 dB in the E-plane.

When excited at the first difference port (in this case, the H-plane difference port), the optimal amplitude balance between the two difference beams was measured as 2.12 dB at 95.2 GHz. Moreover, the lowest null depth was measured as −45.81 dB at 93.8 GHz. The second (E-plane) difference port excitation produced a maximum null depth of −37.71 dB at 95.2 GHz, with an optimal amplitude balance of 0.4 dB achieved at 93.6 GHz. While the slot array is by definition a leaky-wave antenna, the substrate integration with the combiner and coupler network creates an integrated antenna system. Such antennas are the focal point of the next chapter.

The half-mode substrate integrated waveguide has been used as a feed method for transverse slot arrays, and reported by Wu et al. [30]. Figure 2.12 depicts a half-mode substrate integrated waveguide.

This guide was first proposed by Wu et al.—the antenna work followed soon after—and it is advantageous because it retains all the characteristics of the full-mode substrate integrated waveguide, while occupying half the space [31]. For their experimentation process, two transverse slot array antennas were built for two different frequencies. The first design was for a center frequency of 9.8 GHz, while the second was for 31.4 GHz. Both arrays consisted of a total of eight slots and were fabricated on a low-loss 0.508 mm thick substrate. The antennas were fairly narrowband, and measurements of the first antenna revealed a 2.8 % impedance bandwidth, and 2.7 % for the second. The measured 3 dB beamwidth in the

E-plane also remained similar between the two, with 23.3° for the low-frequency version and 21.0° for the high-frequency version.

Work done by Lai et al. further demonstrated the use of the half-mode substrate integrated waveguide, in this instance to feed a dielectric resonator antenna [32]. The energy is coupled from the waveguide to the dielectric resonator through an aperture that separates the two. Experimentation revealed an impedance bandwidth of 24.2 % (for $S_{11} < -10$ dB) around 60 GHz, and maximum radiation efficiency of 92 %. Therefore, this antenna covers the entire ISM band at 60 GHz; the measured gain was around 5.5 dB.

2.4 Closing Remarks

As is evident throughout this chapter, despite the state of development of leaky-wave antennas, there is still a lot of room for innovation in terms of extending these antennas into the millimeter-wavelength region. Trends continue to follow substrate integrated techniques, as opposed to conventional waveguides, as well as antenna on-chip systems.

References

1. T. Tamir, A.A. Oliner, Guided complex waves. Part 1: Fields at an interface. Proc. Inst. Electr. Eng. **110**(2), 310 (1963)
2. T. Tamir, A.A. Oliner, Guided complex waves. Part 2: Relation to radiation patterns. Proc. Inst. Electr. Eng. **110**(2), 325 (1963)
3. L. Goldstone, A. Oliner, Leaky-wave antennas I: Rectangular waveguides. IRE Trans. Antennas Propag. **7**(4) (1959)
4. A. Oliner, D. Jackson, Leaky-wave antennas, in *Antenna Engineering Handbook*, 4th edn., ed. by J. Volakis (McGraw-Hill, New York, 1969)
5. C.A. Balanis, Antenna theory: A review. Proc. IEEE **80**(1), 7–23 (1992)
6. D. M. Pozar, Transmission lines and waveguides, in *Microwave Engineering*, 4th edn. (John Wiley & Sons, Inc., Hoboken, New Jersey, 2012)
7. J.L. Gómez-Tornero, A. de la Torre Martínez, D.C. Rebenaque, M. Gugliemi, A. Álvarez-Melcón, Design of tapered leaky-wave antennas in hybrid waveguide-planar technology for millimeter waveband applications. IEEE Trans. Antennas Propag. **53**(8), 2563–2577 (2005)
8. K.S. Yngvesson, D.H. Schaubert, T.L. Korzeniowski, E.L. Kollberg, T. Thungren, M. Johansson, Endfire tapered slot antennas on dielectric substrates. Trans. Antennas Propag. **12**, 1392–1400 (1985)
9. K.S. Yngvesson, Y.-S. Kim, T.L. Korzeniowski, E.L. Kollberg, J.F. Johansson, The tapered slot antenna—a new integrated element for millimeter-wave applications. IEEE Trans. Microw. Theor. Tech. **37**(2), 365–374 (1989)
10. D. Jackson, A. Oliner, Leaky-wave antennas, in *Modern Antenna Handbook*, ed. by C. Balanis (John Wiley & Sons, Inc., New York, 2008)
11. K.L. Klohn, R.E. Horn, H. Jacobs, E. Freibergs, Silicon waveguide frequency scanning linear array antenna. IEEE Trans. Microw. Theor. Tech. **26**(10), 764–773 (1978)

12. T. Yoneyama, S. Fujita, S. Nishida, Insulated nonradiative dielectric waveguide for millimeter-wave integrated circuits, IEEE MTT-S Int. Microw. Symp. Dig. M(11) 1188–1192 (1983)
13. J.L. Gómez-Tornero, F.D. Quesada-Pereira, A. Álvarez-Melcón, Analysis and design of periodic leaky-wave antennas for the millimeter waveband in hybrid waveguide-planar technology. IEEE Trans. Antennas Propag. 53(9), 2834–2842 (2005)
14. M. García-Vigueras, J.L. Gómez-Tornero, G. Goussetis, A.R. Weily, Y.J. Guo, 1D-leaky wave antenna employing parallel-plate waveguide loaded with PRS and HIS. IEEE Trans. Antennas Propag. 59(10), 3687–3694 (2011)
15. A.J. Martinez-Ros, J.L. Gómez-Tornero, G. Goussetis, Conformal tapered microstrip leaky-wave antennas, Proceedings. of 6th European Conference in Antennas Propagation (EuCAP), no. 2, pp. 154–158, 2012
16. F.K. Schwering, Millimeter wave antennas. Proc. IEEE 80(1), 92–102 (1992)
17. D. Sánchez-Escuderos, M. Ferrando-Bataller, A. Berenguer, J.I. Herranz, Circularly-polarized periodic leaky-wave antenna at millimeter-wave frequencies, in Antennas and Propagation Society International Symposium, 2013 pp. 158–159
18. D. Sanchez-Escuderos, M. Ferrando-Bataller, J. Herranz, and V. Rodrigo-Penarrocha, Low-loss circularly polarized periodic leaky-wave antenna. IEEE Antennas Wirel. Propag. Lett. 1225(c) 1–1 (2015)
19. Y.B. Nechaev, D.N. Borisov, A.I. Klimov, I.V Peshkov, Planar center-fed leaky-wave antenna arrays for millimeter wave systems, in International Conference on Antenna Theory and Techniques, 2015, pp. 1–3
20. T. Zhao, D.R. Jackson, J.T. Williams, A.A. Oliner, General formulas for 2-d leaky-wave antennas. IEEE Trans. Antennas Propag. 53(11) 3525–3533 (2005)
21. T.Z.T. Zhao, D.R. Jackson, J.T. Williams, H.-Y.D. Yang, A.A. Oliner, 2-D periodic leaky-wave antennas—Part I: Metal patch design. IEEE Trans. Antennas Propag. 53(11), 3505–3514 (2005)
22. D.R. Jackson, A.A. Oliner, Leaky-wave propagation and radiation for a narrow-beam multiple-layer dielectric structure. IEEE Trans. Antennas Propag. 41(3), 344–348 (1993)
23. H. Yang, N. Alexopoulos, Gain enhancement methods for printed circuit antennas through multiple superstrates. IEEE Trans. Antennas Propag. 35(7), 860–863 (1987)
24. D. Jackson, N.G. Alexopoulos, Gain enhancement methods for printed circuit antennas. IEEE Trans. Antennas Propag. APP-33(9) 976–987 (1985)
25. K. Wu, Y.J. Cheng, T. Djerafi, W. Hong, Substrate-integrated millimeter-wave and terahertz antenna technology. Proc. IEEE 100(7), 2219–2232 (2012)
26. T. Zhao, D.R. Jackson, J.T. Williams, 2-D periodic leaky-wave antennas—Part II: Slot design. IEEE Trans. Antennas Propag. 53(11), 3515–3524 (2005)
27. F. Xu, K. Wu, X. Zhang, Periodic leaky-wave antenna for millimeter wave applications based on substrate integrated waveguide. IEEE Trans. Antennas Propag. 58(2), 340–347 (2010)
28. Y.K. Fuh, A. Margomenos, Y. Jiang, L. Lin, Micromachined W-B band plastic slot array antenna with self-aligned and integrated Flange, 15th International Conference in Solid-State Sensors, Actuators, Microsystems, vol 1 (2009) pp. 2122–2125
29. Y.J. Cheng, W. Hong, K. Wu, 94 GHz substrate integrated monopulse antenna array. IEEE Trans. Antennas Propag. 60(1), 121–129 (2012)
30. J. Xu, W. Hong, H. Tang, Z. Kuai, K. Wu, Half-Mode substrate integrated waveguide (HMSIW) leaky-wave antenna for millimeter-wave applications. IEEE Antennas Wirel. Propag. Lett. 7, 85–88 (2008)
31. W. Hong, B. Liu, Y. Wang, Q. Lai, H. Tang, X.X. Yin, Y.D. Dong, Y. Zhang, K. Wu, Half mode substrate integrated waveguide: A new guided wave structure for microwave and millimeter wave application, Joint 31st International Conference in Infrared Millimeter Waves and 14th International Conference in Terahertz Electron. (IRMMW-THz), vol 152 (2006) p. 219
32. Q. Lai, C. Fumeaux, W. Hong, R. Vahldieck, 60 GHz aperture-coupled dielectric resonator antennas fed by a half-mode substrate integrated waveguide. IEEE Trans. Antennas Propag. 58 (6), 1856–1864 (2010)

Chapter 3
Printed and Planar Antennas

Many applications of RF systems impose severe space constraints on the design of the antenna. Military aircraft, satellites, guided missiles, and mobile broadband systems (to name but a few) need low-profile antennas that effortlessly integrate into a mechanical structure. One class of antennas that may be used to address this problem is planar antennas, where all the elements are in the same plane. These antennas are low-profile, generally very simple to manufacture and well geared toward photolithographic methods. Moreover, they have the potential to sustain a lot of mechanical stress when they are mounted on rigid surfaces. In terms of polarization, antenna pattern, and resonant frequency, these antennas are very versatile and can be designed for a wide array of requirements. Serious limitations often associated with printed antennas, however, are their low power-handling capability, low efficiency, and very narrow achievable bandwidths, although there are many techniques that have been thoroughly investigated that yield higher bandwidths. At millimeter wavelengths, these antennas are often built on top of thick substrates with high permittivity.

Surface waves and mutual coupling present significant challenges to the designer because of their degradation of the antenna performance, and their effects become more obvious as the height of the substrate is increased. Characteristics such as bandwidth and input impedance are also heavily influenced by the substrate thickness and the approach to feeding methods and dealing with dielectric losses is quite different from dealing with lower frequencies.

Two types of printed antennas are extensively used in practice: microstrip elements and printed dipoles, although it can be argued that printed dipoles are simply a subset of microstrip planars. Patch elements come in just about any shape and size, and are rarely implemented in single-element configurations. As a result, the focus of this chapter will primarily remain on array configurations and variations thereof. Other printed antennas include loops, slots, Yagi-Uda antennas, and planar inverted-F antennas. It should also be mentioned that there is a distinction to be made between on-chip and off-chip printed antennas. On-chip integrated antennas will be covered in detail in a later chapter.

© Springer International Publishing Switzerland 2016   39
J. du Preez and S. Sinha, *Millimeter-Wave Antennas: Configurations and Applications*, Signals and Communication Technology,
DOI 10.1007/978-3-319-35068-4_3

3.1 Classification of Printed and Planar Antennas

As antenna technology evolved over the years, it was expected that each subset would contain a multitude of varying shapes, excitations, and configurations. This is no different for planar antennas. In order to review the state of the art in printed planar antenna technology effectively, this chapter is separated into several sub-classes that should ease the process of following trends between the different antennas. It is worth noting that some of the antennas discussed in the previous chapter are, strictly speaking, planar or printed (or both); the principles on which they operate are quite different from those of the antennas that will be discussed here. The simplest printed antenna (and one of the most well-known) is the rectangular microstrip element. Many printed configurations are derived from this simple structure.

3.2 Microstrip Elements

The idea of microstrip antennas can be traced back to the early 1950s [1], although the interest of the technical community it enjoys today was not generated until the 1970s. Microstrip antennas are often chosen for an application for their mechanical structure rather than their electrical performance. These antennas are extremely compact and arrays of thousands of elements can be fit onto reasonably sized panels.

Microstrip patches are designed to radiate in the broadside direction, in such a manner that the main beam will be orthogonal to the patch element. A simple illustration of a rectangular patch antenna is shown in Fig. 3.1.

The antenna in Fig. 3.1 is reminiscent of the early stages of development that microstrip antennas underwent, when the approach was typically to design and etch a rectangular metallic patch on top of an electrically thin substrate. The bottom layer of the substrate was covered with a ground plane, and the patch element itself

Fig. 3.1 Rectangular
microstrip antenna, shown as
a radiating patch on a
grounded dielectric substrate

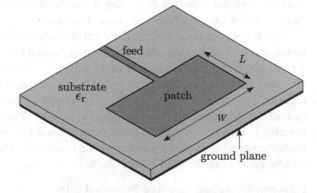

was usually made from thin copper (around 35 μm). Since then, a multitude of shapes other than the rectangle have been used to create individual elements.

3.2.1   Theory of Operation

Several methods are known that can be used to analyze microstrip antennas. Some of these are the transmission line model, the cavity model, or a full-wave analysis [2–5]. The transmission line model is the simplest, but it also returns the least accurate results [6]. The transmission line equivalent of a rectangular microstrip is shown in Fig. 3.2 [7]. The patch consists of two radiating slots, each with an equivalent admittance given by $Y = G + jB$.

Despite its inaccuracy, the transmission line model is nonetheless a useful tool, and many microstrip element designs begin by using this model because of its simplicity. Further optimization and modeling is then done in an appropriate computer-aided design package. That said, designing a microstrip antenna with a rigorous method such as the aforementioned full-wave analysis requires knowledge of the approximate dimensions of the antenna. These can either be derived from closed form equations or will simply be known to the designer from experience.

The concept of an effective dielectric constant is an extremely important approximation to the operation of microstrip lines. Consider the microstrip line shown in Fig. 3.3. If the dielectric substrate were absent, the configuration in Fig. 3.3 would simply be a two-wire transmission line that is inside a homogeneous medium, in this case air. Given this arrangement, one would be dealing with a simple TEM line with a known propagation constant and phase velocity. However, the inclusion of a dielectric substrate in this configuration complicates matters, even more so since the dielectric does not fill the region above the microstrip line.

Toward the edges of the microstrip line, the electric fields are no longer as tightly coupled to the ground plane and a condition arises where some of the fields are contained within the dielectric, while some are contained in the air. Waves traveling in the dielectric would not have the same phase constant as waves traveling in air, which means that it is not possible to enforce a phase-matching condition between

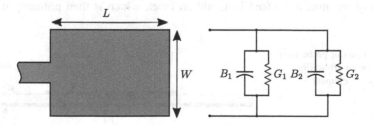

Fig. 3.2 Rectangular microstrip antenna shown alongside its equivalent circuit, based on the transmission line model

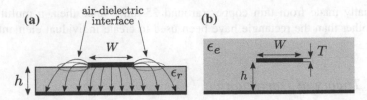

Fig. 3.3 Microstrip transmission line with **a** electric field lines and air–dielectric boundary and **b** effective dielectric constant configuration

the two media. For this reason particularly, a microstrip line cannot support a pure TEM wave [8, 9]. For most applications, however, the substrate is electrically thin ($h \ll \lambda$), the fields are quasi-TEM and the effective dielectric constant can be computed by

$$\epsilon_e = \frac{\epsilon_r + 1}{2} + \frac{\epsilon_r - 1}{2}\left(1 + \frac{12h}{W}\right)^{-1/2}. \tag{3.1}$$

This serves as an effective approximation. The choice of substrate is a critical part of extending the frequency of operation into the millimeter region, as will become clear throughout this chapter.

3.2.2   Feeding Methods

To feed the antenna, four configurations are most often implemented in practice, namely the coaxial probe, microstrip transmission line, aperture coupling, and proximity coupling [10]. The microstrip transmission line was shown in Fig. 3.1, and the coaxial probe method is shown in Fig. 3.4.

The coaxial feed method is highly dependent on the substrate thickness; this should be minimized in order to minimize parasitic inductance added by the connector pin, although it produces minimal spurious radiation. On the other hand, using a microstrip feed allows the designer to control the input impedance to the radiating element easily and it is extremely simple to fabricate. A popular modification of the microstrip feed is to add an inset, which is then primarily used to

Fig. 3.4 Coaxial probe feed method for an arbitrarily shaped patch element

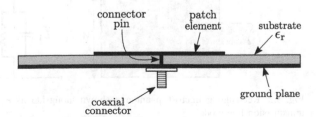

Fig. 3.5 Inset-feed method

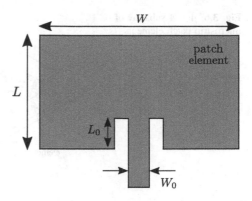

control the impedance. Such a feed is shown in Fig. 3.5, where the inset length, line width, and spacing can be modified to obtain the desired impedance at a particular center frequency.

A disadvantage of the microstrip and coaxial feed methods is that both introduce asymmetries into the radiating element, which excite higher order modes and result in higher cross-polarization [11].

Using aperture-coupled and proximity-coupled feeds are two viable alternatives that overcome some of the issues associated with microstrip and coaxial feeds. In an aperture-coupling feed, the coupling is achieved by separating two substrates with a ground plane. Underneath the bottom substrate is a microstrip transmission line and the energy is coupled through a slot in the ground plane onto the patch. One advantage of this approach is that it enables independent optimization of the radiating element and the feeding mechanism, but it is more complicated to fabricate in comparison to other methods.

The proximity-coupling method presents the largest bandwidth of about 13 % of the four feeds discussed here, but it is also subject to a more complicated fabrication process, which can be especially challenging in the millimeter-wavelength region.

Large arrays of microstrip elements are often fed through a corporate feed network, although it is also possible to feed these elements in series. Such series-fed arrays are limited to having a fixed beam, unless some type of frequency scanning is implemented. Another disadvantage is that nonidealities in one element affects the remainder of the elements in the series, and this is not entirely the case in a corporate feed. An example of a corporate feed is shown in Fig. 3.6.

The subarray shown here is part of a larger array. It is commonplace in practice to approach large arrays in this manner because complex feed networks can be abstracted into smaller sections that are somewhat easier to design [12]. In this example, beam scanning is achieved by adding variable phase shifters before the transmission lines that feed each subarray. It is also possible to include an appropriate pattern tapering by altering the track impedance of individual branches in the network, although it should be ensured that each element in a particular subarray is driven in phase.

Fig. 3.6 4 × 4 subarray of
microstrip patch elements

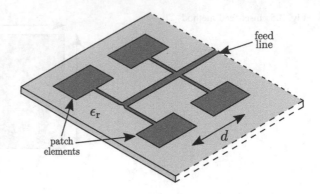

3.2.3 Design Principles

Patch elements are extremely flexible when it comes to their physical shape and can
often be used to tailor the antenna to a specific application. Balanis has outlined a
set of design guidelines, based on the transmission line model, which quickly leads
to a first iteration of the patch element that can then be tweaked and optimized [11].
While it is not possible (nor is it the intention of this text) to provide a detailed
design guideline on various types of printed antennas, it is nevertheless included to
provide context on many of the discussions on the design of these antennas in
succeeding sections.

The element dimensions are finite along its length and width, and as a result, the
fields at the extremities of the microstrip suffer from fringing. The extent to which
this fringing occurs depends on both the height of the substrate and the dimensions
of the patch. We will not go into all the details of fringing here, but it is an
important consideration in patch design because it causes the element to appear
electrically larger than it actually is. This ties into the approximation that is used in
the transmission line model; since the phase discontinuity caused by the air–dielectric interface means that a microstrip cannot support a pure TEM wave, as
discussed in detail in an earlier section.

For a rectangular microstrip element excited in the fundamental mode, the center
frequency is given by

$$f_r = \frac{c_0}{2(L+h)\sqrt{\epsilon_e}}. \qquad (3.2)$$

In (3.1) and (3.2), L and W indicate the resonant length and width of the patch,
respectively, while h indicates the substrate height and ϵ_r its dielectric constant. The
free space velocity of light is indicated by c_0 and 3×10^8 m/s is commonly used,
although this is not its exact value. It should be noted that (3.2) is valid only if the
W/h ratio is greater than unity. The values for L and W, however, are unknown at
this point, and one can begin by finding the width as

$$W = \frac{c_0}{2f_r}\sqrt{\frac{2}{\epsilon_r + 1}}. \tag{3.3}$$

This width is a decent approximation that leads to an acceptable radiation efficiency. Once W is known, the effective dielectric constant can be found by evaluating (3.2) and the extension in length that results from the fringing effect is then given by

$$\Delta L = 0.412h\frac{(\epsilon_e + 0.3)(\frac{W}{h} + 0.264)}{(\epsilon_e - 0.258)(\frac{W}{h} + 0.8)}. \tag{3.4}$$

The actual length of the patch can then be determined by evaluating

$$L = \frac{1}{2f_r\sqrt{\epsilon_e}\sqrt{\epsilon_0\mu_0}} - 2\Delta L. \tag{3.5}$$

As discussed earlier, the equations used here yield a basic element with approximately the correct dimensions. The model can then be optimized and tuned based on the desired feeding method and array configuration.

3.2.4 Considerations for Millimeter-Wave Operation

As would be expected, extending microstrip antennas into millimeter wavelengths is not a simple matter of scaling the structure with wavelength. Two challenges in particular become critically obvious at millimeter wavelengths, namely losses in the feed networks and tolerances associated with fabrication. Typically, arrays that operate in the millimeter band requiring transmission lines in the feed network have lengths that are around one tenth of a millimeter. This imposes significant tolerances on the fabrication and increases the cost of the process, seeing that more expensive equipment is generally required. Furthermore, to obtain a highly directive pattern, many elements are often placed together in an array. Microstrip lines are not low-loss in nature and large corporate feed networks begin to introduce significant losses. As a result, improvement of the efficiency of these antennas has been intensely studied in the last few decades.

The bandwidth limitation can be approached by using an electrically thick substrate. If the antenna is operated close to the ground plane, as would be the case for an electrically thin substrate at microwave frequencies, the use of resonator antennas is required to raise the radiation resistance [13]. As the frequency increases, the thickness of the substrate in wavelengths increases and it may thus be possible to have a substrate that is, e.g., $\lambda/4$ thick while being physically thin. The result is that it is no longer necessary to use high-Q resonator antennas, but rather some other configuration that provides a larger bandwidth. Thicker substrates also serve to mitigate the fabrication tolerances somewhat.

On the other hand, a thick substrate tends to trap surface waves, which do not add to the desired radiation pattern, and instead their radiated power is considered antenna loss. Surface waves can also serve to distort the radiation pattern, since they will radiate when they are perturbed while propagating in the substrate. This parasitic radiation reduces the antenna efficiency. The bandwidth/efficiency problem was thoroughly investigated (independently) by Pozar [14] and the Alexopoulos group [15], and their results were in agreement. A good balance is achieved with a substrate permittivity of less than 4.5 and height of about a quarter of a wavelength.

The Alexopoulos group focused heavily on printed dipoles and with the prescribed substrates the radiation efficiency exceeded 50 %, while the achievable bandwidth with wider dipoles is expected to be greater than 18 %. Although moderately low, this is still a significant improvement on the attainable bandwidth with linear dipoles.

As discussed in the earlier section, large arrays (often consisting of printed dipoles) are required to obtain highly directive radiation patterns. This introduces new challenges in antenna design. Mutual coupling between elements is already a serious concern and this is exacerbated by the propagation of surface waves in the substrate. The requirement is thus clear for a low-loss feed network that is simple to fabricate and that minimizes radiation in the feed line. One of the early solutions to the feed network question is a two-layer substrate proposed by Katehi and Alexopoulos [16]. The feed network is printed as microstrip lines on the bottom substrate. This substrate is chosen to be electrically thin, which means that the waves traveling on these lines are tightly bound, resulting in a lower overall level of feed system radiation. The antennas (in this example, microstrip dipoles) are printed on the top layer, and the excitation comes from the coupling to the bottom layer network.

3.3 Research Review

Given the fact that microstrip antennas are very rarely used in single-element configurations in practice, a substantial portion of research effort is aimed at optimizing arrays of patch elements, rather than proposing new approaches to element design. This is not to say that these proposals do not exist, and on the contrary a large number of interesting concepts are still introduced every year. The extension of these elements into arrays is inevitable, and this section will cover some of these developments.

Two methods that are most often encountered in wideband applications is using either a dipole or a slot of a quarter-wavelength and placing the element above a ground plane. The problem, however, is that the variation with frequency that the electrical height of the antenna experiences causes the radiation pattern to vary greatly over the specified bandwidth. This variation can of course be within acceptable limits for some applications, but generally it is not desirable and therefore should be accounted for. Another issue is high levels of cross-polarization, more so in the higher frequency portion of the desired spectrum. Several methods have been proposed over the years to reduce cross-polarization, such as W-shaped ground planes [17], antiphase cancelation [18], and meandered feeding methods [19], although gain and beamwidth variations remain an issue.

Another typical measure of performance in wideband antennas is the input reflection coefficient (S_{11}) over a specified bandwidth, often referred to as the impedance bandwidth.

In recent years, printed antennas in a multitude of configurations have been proposed and evaluated by several authors and we will attempt to cover as many of these as possible. Much of the effort has been aimed at bandwidth-enhancing techniques, as will become evident as the chapter moves forward.

3.3.1 U-Slot Microstrip

Bandwidth enhancement is a major concern in the design of microstrip antennas. In the last few decades, a multitude of techniques have been proposed to increase the impedance bandwidth of microstrip antennas in the low-GHz range. As the frequency increases into the millimeter-wave region, the difficulties in antenna design move away from structural challenges to the fabrication process. Given the tolerances that have to be contended with at millimeter wavelengths, antennas that are mechanically complex are simply unfeasible. The U-slot loaded patch is an antenna with a simple structure that has been demonstrated to work on a single-layer substrate. A basic version of a U-slot microstrip antenna element is shown in Fig. 3.7.

A similar configuration to the one shown here was first introduced in 1995 by Huynh and Lee [20], described as a single-layer, single-element wideband patch

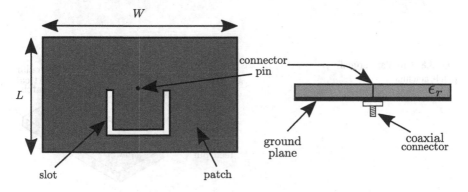

Fig. 3.7 Pin-fed microstrip antenna with a U-shaped slot inserted, placed on a grounded substrate

antenna with linear polarization. At RF and microwave frequencies, a number of studies reported on the obtainable impedance bandwidth of the U-slot patch antenna and the general consensus was that over 30 % is achievable in air substrates, and about 20 % for material substrates [21]. In both cases, the substrate thickness was about 0.08 λ_0.

One proposal for such a U-slot patch was reported by Sun et al. [22] in 2013, and the antenna was specified in accordance with the IEEE 802.15.3c standard for 60 GHz networks [23]. The standard describes the division of about 9 GHz of spectrum into four 2.16 GHz channels, from 57.24 GHz up to 65.88 GHz, therefore imposing a requirement for a wide impedance bandwidth in antennas. Furthermore, a gain of approximately 15 dBi is required for indoor wireless networking applications and circular polarization is especially desirable in order to reduce the effects of multipath [23]. To obtain circular polarization, the corners of the microstrip element can be truncated, or alternatively, the U-slot can be made asymmetric.

The proposal from Sun, Guo and Wang was one of the earliest investigations on U-slot patch antennas at millimeter-wave frequencies. The antenna is fabricated on low-temperature cofired ceramic technology, mainly because of the ability to insert high-quality passive components in the ceramic while still supporting the addition of active devices. The final antenna was fabricated into a 4 × 4 array, totaling 14 × 16 × 1.1 mm^3 in size.

3.3.2   *Vertical Patch*

The vertical microstrip antenna has been extensively investigated and reported on for microwave frequencies [24–27]. This microstrip configuration possesses several advantages, such as being mechanically simple, having a very wide impedance bandwidth and exhibiting a directional radiation pattern that is reasonably stable over its bandwidth. As expected, the millimeter-wave implementation of the vertical patch antenna poses several new challenges. A conceptual drawing of a circular vertical patch is shown in Fig. 3.8.

Fig. 3.8 Circular vertical patch antenna

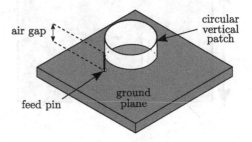

air gap

circular vertical patch

feed pin   ground plane

The patch element itself becomes extremely small at millimeter-wave frequencies, increasing the requirement for sophisticated fabrication, and it also requires precise alignment to the ground plane during the assembly procedure.

One approach to solving these difficulties is using a vertical radiating metallic ring that is embedded into a substrate, proposed by Wong et al. [28]. The specified bandwidth stretches from 28.4 GHz to 37 GHz, aligning this antenna with several K_a-band applications. The antenna is constructed by inserting a vertical circular patch into a dielectric substrate and attaching a microstrip ring to the base of the vertical element. To control the input impedance, the parameters of the microstrip ring can be altered and it is connected to a conductor-backed coplanar waveguide feed. The antenna consists of two stacked dielectric layers; the grounded coplanar waveguide is located on the bottom layer and the microstrip ring on the top layer.

On the results side of things, the impedance bandwidth (for $S_{11} \leq 10$ dB) was measured as 26 %, and the broadside gain was reported as varying between 7 and 8.7 dBi. Over the specified bandwidth, the E-plane and H-plane radiation patterns were relatively stable, and the antenna exhibited a cross-polarization of 20 dB. No mention was made of the sidelobe level. The measured E-plane pattern at 33 GHz presented a deep null at approximately 45° off broadside, which was not the case in simulation. However, the authors did mention a problematic flange mount with the SMA connector, which proved to radiate spuriously and was likely to be the source of the null.

3.3.3 Magneto-Electric Dipole

Magneto-electric dipoles are wideband, unidirectional antennas that are built from a planar electric dipole together with a vertically positioned magnetic dipole. The vertical dipole is formed with a shorted patch antenna. Although this is the general approach at lower frequencies up to the microwave range, it becomes impractical at millimeter wavelengths and another approach should be followed.

Since the original proposal of this antenna by Luk and Wong [29, 30], a large number of studies have followed. A dual-polarized version tailored toward mobile communications [31], a circularly polarized variant for use in satellite broadcasting [32] and an ultra-wideband version for wireless networking applications [33] are some prime examples. Unfortunately, none of these variations operate at millimeter-wave frequencies and for several years there have been no feasible suggestions for millimeter-wave operation of magneto-electric dipoles.

The requirement for low-cost, wideband antennas in the millimeter-wave region steadily increases as 60 GHz networks become a reality. Integrated antennas for millimeter waves are extremely popular, mainly because of their footprint size and their capability to fit easily into existing systems. Most of these on-chip antennas do, however, suffer from very low achievable gains.

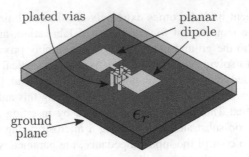

Fig. 3.9 Conceptual magneto-electric dipole suited for millimeter-wavelength operation

The magneto-electric dipole proposed later on by Ng et al. is formed through plated vias that are shorted to the ground plane [34]. A conceptual drawing of such an antenna is shown in Fig. 3.9. The reported antenna used a slightly different structure on the top layer, and the feed was realized through a coupled T-shaped strip.

The coupled T-shaped strip is one of many approaches to feeding a magneto-electric dipole and is shown among some other viable techniques in Fig. 3.10.

For the feeding methods shown in Fig. 3.10, the T-shaped strip offered the largest bandwidth. The direct connection offered approximately 14 % impedance bandwidth, while the L and T-shaped feeds resulted in over 40 %. In terms of cross-polarization, the direct feed exhibited the lowest levels for both the E-plane and H-plane measurements. The T- and L-shaped coupled feeds exhibited higher levels of cross-polarization, the highest being measured for the L-shaped case, which was probably due to the asymmetric current that flowed in the coupled strip. Following these results, it can be declared that the T-shaped feed is the optimal choice for wideband applications, while the direct feed is suitable for narrowband.

The overall performance of the antenna reported in [34] was analyzed using the T-shaped coupled feed. It had a measured impedance bandwidth of 33 % (for $S_{11} < -15$ dB) over an ultra-wide bandwidth of 20 GHz, centered at 60 GHz. Over

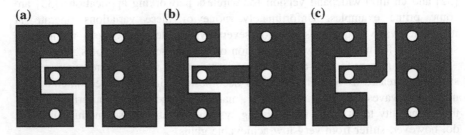

Fig. 3.10 Three feed methods typically used in magneto-electric dipoles: **a** the coupled T-shape strip, **b** the direct feed, and **c** the coupled L-shaped strip

the same bandwidth, the cross-polarization peaked at −15 dB and the authors deemed the radiation pattern adequately stable. The measured gain was found to be reasonably consistent, varying between 6 and 8 dB over the specified bandwidth. The plated through holes (vias) used to realize the magnetic dipole are extremely inexpensive; coupled with the simple microstrip shapes, this makes this antenna reasonably cheap to fabricate.

This novel approach to magneto-electric dipole implementation has led to a number of publications that investigate the approach and propose alternative feeding methods and patch configurations. The first of these (also reported on by Ng et al.) that we will be discussing is essentially a coplanar-waveguide-fed slot variation of the earlier magneto-electric dipole [35]. The idea is to improve on the narrow bandwidth offered by a traditional coplanar-waveguide-fed slot antenna by implementing a magnetic source as well as a parasitic dipole, which acts as an electric source. To elucidate this discussion, a simple comparison between a planar slot antenna and one with electric dipoles is shown in Fig. 3.11.

The antenna is differentially fed, relieving some of the difficulty of integrating it alongside differential power amplifiers and low-noise amplifiers. Unlike the earlier antenna that used planar dipoles to realize the electric source and plated parallel vias to realize the magnetic source, this antenna forms the electric source through parasitic dipoles on the ends of a slot, which serve as the magnetic source. The complementary source is implemented in a stacked configuration by using two substrates, one for each of the sources. In the experimentation carried out by Ng et al., both of the dielectric layers were exactly the same substrate (Rogers RT/Duroid 5880, with dielectric constant of 2.2, and a thickness of 30 mils). The plated vias were 0.2 mm in diameter for the proposed structure.

The planar dipoles at the ends of the slot are printed on the top substrate, and they are differentially excited with equal amplitudes through the plated vias. The slot, which acts as the magnetic source in this configuration, is printed on the bottom layer substrate and is fed by a coplanar waveguide on the same layer. Furthermore, the coplanar waveguide is then connected through three vertical pins in a ground-signal-ground configuration. The simple element using an oval-shaped dipole was used in a 2 × 1 array configuration, as illustrated in Fig. 3.12.

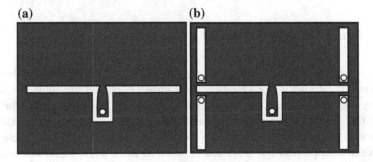

Fig. 3.11 **a** Traditional slot antenna and **b** with thin dipoles added to the structure

Fig. 3.12 2 × 1 array
configuration of a
complementary source slot
antenna

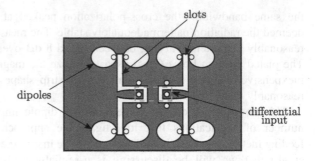

This configuration yielded improved cross-polarization levels of 23–25 dB at 60 GHz and 25 dB at 67 GHz. An impedance bandwidth of just under 20 % was achieved (for a VSWR ≤ 3:1), and the measured gain peaked at 12.2 dBi, while remaining above 10 dBi from 55 to 77 GHz. One limitation in the reported results was the lack of a 60 GHz differential source, which required a waveguide-to-microstrip antipodal finline transition to provide the differential excitation. It is likely that measurements obtained from a 60 GHz differential source would have been somewhat different, since there would be one less nonideality to deal with. Nonetheless, the results and concepts reported in this paper are invaluable.

Li and Luk [36] recently proposed a wideband and circularly polarized version of the magneto-electric dipole. Circular polarization was obtained by altering the dimensions of a linearly polarized magneto-electric dipole. A low-frequency prototype was first constructed, and this version achieved a 73.3 % impedance bandwidth around 2 GHz, with a gain of 6.8 dBi and a 3 dB axial ratio of 47.7 %. This design was then extended into a millimeter-wave antenna operating around 60 GHz. The prototype version was implemented on a single-layer dielectric substrate and achieved a 56.7 % impedance bandwidth, along with a maximum gain of 9.9 dBi and a 3 dB axial ratio of 41 %.

The wide bandwidth achieved by this antenna is largely due to the L-shaped probe feed configuration and the conventional printed circuit technique used to etch the antenna is highly advantageous because of its mainstream use and low fabrication cost.

3.3.4 Planar Yagi-Uda Arrays

The Yagi-Uda antenna was first introduced in 1927 at the University of Japan by Professor Uda. Roughly one year later, an English paper on the subject was published by one of Professor Uda's colleagues, Yagi [37]. Both described the operation of this newly invented antenna, and while Yagi cited the work of Uda, the antenna was called a Yagi antenna. In order to acknowledge the contributions of both inventors appropriately, it should be referred to as a Yagi-Uda antenna.

Fig. 3.13 Basic
configuration of a Yagi-Uda
antenna. The elements are
arranged in a linear array in
the y-direction

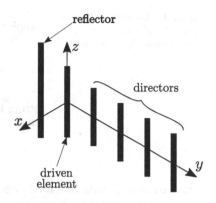

The Yagi-Uda antenna is a traveling wave antenna, which is formed by placing linear dipoles in an array configuration. This is illustrated in Fig. 3.13. The antenna can be divided into three parts, as shown in the figure. These are the arrangement of the reflector, the feeding element, and the rows of directors.

One interesting development that has surfaced in recent years is the concept of a quasi-Yagi antenna. The Yagi-Uda array is a very popular broadband traveling wave antenna at frequency bands from HF to ultra high frequency (UHF) (3 MHz–3 GHz), but designers have had limited success in extending the principles into the microwave and millimeter-wave region. Some of the solutions that have been proposed to operate these antennas in the millimeter-wave region is a planar microstrip Yagi [38], and a coplanar Yagi-Uda array, which is fed by a stripline and uses a reflector element that is printed on a thick dielectric slab [39]. A broadband quasi-Yagi antenna, realized on a substrate with a high dielectric constant, was first introduced by Kaneda et al. [40]. The proposed configuration was easily integrated with microstrip circuitry, since the driver was fed by a microstrip line. The antenna yielded a large bandwidth of 48 % for a VSWR below 2:1, as well as cross-polarization levels of less than −15 dB at X-band.

Alhalabi and Rebeiz [41] proposed a differentially fed Yagi-Uda antenna designed to operate at 22–26 GHz. The design experimented with a number of different dipole feeds, although a much greater impedance bandwidth was obtained from using a folded dipole. The geometry and layout for the planar Yagi-Uda antenna are shown in Fig. 3.14.

The antenna is etched on a thin Rogers RT/Duroid 5880 substrate (0.381 mm) with a low dielectric constant ($\epsilon_r = 2.2$). One advantage of the structure shown in Fig. 3.14 is that it can be directly connected to a differential RF circuit module. To complete the radiation pattern measurements, Alhalabi and Reibez used a lock-in amplifier together with a Schottky diode detector with the antenna in the receive mode. The test signal is amplitude-modulated on a 1 kHz carrier and the rectified signal at the receiver end is measured with the lock-in amplifier. On the other hand, gain measurements were done using a 180° hybrid to feed the antenna through a microstrip line.

At 24 GHz, the measured E-plane 3 dB beamwidth was 42° and the same beamwidth was 70° in the H-plane. Over the bandwidth of 22–26 GHz, the

Fig. 3.14 Geometry and layout of a planar, dipole-driven Yagi-Uda array

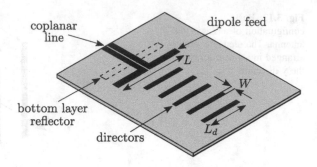

measured cross-polarization levels were below −22 dB. Aside from some variation introduced by the coaxial connector, the measured return loss agreed with the simulation and S_{11} was below −10 dB over the specified bandwidth. The measured gain was found to be at least 8 dB over the bandwidth, and the maximum gain value was 9.8 dB. Future work includes serious investigation of ways to further increase the frequency of this antenna.

Qin et al. [42] reported on a reconfigurable quasi-Yagi antenna in 2010. Similar to the configuration proposed by Alhalabi and Reibez, a folded dipole is used as the driver element. To alternate between two driven elements, two PIN diodes (a microwave diode with p-type, intrinsic and n-type semiconductor layers) are added between two strips in the folded dipole driver. In other words, the driven element is changed between one or two parasitic directors together with the folded dipole. To mount the PIN diodes, a conductive silver epoxy is used to glue them between the gaps between the dipole driver and the primary microstrip director element. The particular antenna discussed here is capable of switching between the 57–60 GHz band and the 71–76 GHz band. One advantage of such a reconfigurable configuration is that the antenna can filter interference signals from bands that are not currently in use, thereby greatly relaxing filtering requirements in the analogue front end. The reconfigurable antenna thus has inherent bandpass characteristics. A feed network consisting of a coplanar stripline and transformer balun facilitates biasing of the PIN diodes. A folded dipole is advantageous in this regard, since a regular half-wavelength dipole would require additional biasing lines, leading to some degree of pattern distortion [43].

This antenna was machined on a thin (0.1 mm) liquid crystal polymer substrate with a permittivity of $\epsilon_r = 3.2$, which is suitable for reasonably low-cost fabrication techniques and presents low losses at millimeter-wave frequencies. The measured bandwidth at the lower band was 26.6 %, while the upper band resulted in an 8.2 % bandwidth for $S_{11} < -10$ dB.

Another millimeter-wave Yagi that uses parasitic patch elements in a two-layer configuration was presented recently by Briqech and Sebak [44], albeit only in simulation. Both layers of the structure consist of a driven element, a reflector, and two director elements. The array on the first layer is driven from a source and in turn excites the array on the second layer. This antenna has been shown to be capable of very high radiation efficiencies, exceeding 95 % with a wide impedance bandwidth of 19.4 %, covering the entire Industrial, Scientific and Medical (ISM) band at 60 GHz.

Stacking multiple layers of Yagi antennas is an effective method to reduce the overall footprint of the structure. Another configuration, proposed by Kramer, Djerafi and Wu [45], consists of a 4 × 4 array of Yagi-Uda antennas that measures 28 × 24 × 2.4 mm in size. The idea of stacking the antennas is based on overcoming the gain saturation experienced by planar arrays by introducing a third dimension. A vertical patch element is used as the driver and there are four parasitic director elements. In this design, all the patch elements are circular, but it is not limited to circular elements. Using circular elements is advantageous, because of the possibility of a dual-polarized configuration and the simplicity of optimizing the element, since only the radius is varied in the process.

The antenna consists of a total of six layers of the Rogers RT/Duroid 5880 substrate material, with a dielectric constant of $\epsilon_r = 2.2$. Each layer is 0.508 mm thick. To increase the bandwidth, the driven element is fed through an aperture-coupled mechanism. In their theoretical analysis, the authors concluded that increasing the number of elements leads to a roughly linear increase in gain, but saturation begins to occur at around eight or nine elements. The gain value at this point is approximately 14 dB. In comparison to using the dipole elements, circular patches provide higher achievable gain and the reflector (the ground plane below the patch element) is notably larger. Furthermore, the air–dielectric interface surrounds the antenna completely, leading to reduced levels of spurious radiation and very low levels of dielectric loss (at most 0.5 dB). For millimeter-wave circuits integrated on the substrate, this could be particularly advantageous.

The authors optimized the stacked Yagi-Uda antenna as a single element and then proceeded to construct a 4 × 4 array, fed by a substrate-integrated waveguide network. The substrate-integrated waveguide is extremely suitable for millimeter-wave circuits owing to low radiation, low loss, and ease of integration. To realize the aperture coupling to the circular patch, the top layer of the feed network contains electroplated slots (vias) as driver elements. This feed configuration, however, does not permit the use of a reflector.

The measured impedance bandwidth of the array was 7 % (for $S_{11} < -10$ dB), and the antenna had a gain of 18 dBi. On the other hand, the single element resulted in a gain of 11 dBi over a 4.2 % bandwidth, with a 6.5 × 6.5 × 3.4 mm footprint. An improvement in sidelobe level of approximately 6 dB was achieved by angling the antenna elements, with negligible gain reduction. A microwave implementation of their stacked Yagi-Uda antenna with dual polarization was reported before the authors explored millimeter-wave possibilities [46].

3.3.5 Conventional Printed Circuit Antennas

The use of low-temperature co-fired ceramic technology has enabled the construction of complex multilayer patch antenna configurations. On the other hand, the conventional printed circuit technique is advantageous if the antenna can be designed as a single-layer structure. Printed circuit technology is realizable on relatively cheap materials through a simple manufacturing process, although the narrow achievable impedance bandwidth (typically around 5 %) offered by single-layer antennas at millimeter wavelengths is a major limitation. In order to overcome this limitation, a novel coplanar feed network was recently proposed by Li and Luk [47]. The microstrip element is proximity-coupled to an L-shaped probe and this probe is realized as a plated via in the substrate together with a microstrip line. Therefore, the section consisting of the radiating patch and the microstrip portion form an open-circuited stub at the end of a coplanar waveguide and this section is responsible for the wideband characteristics of this antenna.

This element was expanded into two 4 × 4 arrays, one with linear polarization and the other with circular polarization. As discussed earlier, circular polarization can be a desirable attribute for short-range communication links where multipath interference poses a significant challenge. The two configurations used the same coplanar feed network, and in order to obtain circular polarization, the antenna elements were sequentially rotated throughout the array.

The antenna exhibited excellent wideband characteristics, with a 25.5 % impedance bandwidth (for a VSWR less than 2:1) and a gain of 15.2 dBi in the linearly polarized version, and a 17.8 % impedance bandwidth and a gain of 14.5 dB for the circularly polarized version. Furthermore, the circularly polarized variant achieved a 3 dB axial ratio of 15.6 %.

3.3.6 Micromachined Patch Antennas

Throughout this chapter, we have emphasized the fabrication difficulties that are commonly associated with printed antennas at millimeter wavelengths. Silicon and gallium arsenide substrates typically used have very high dielectric constants (e.g. $\epsilon_r = 11.7$ for silicon), and as a result, surface waves are easily excited in the substrate. To reduce the power radiated from surface waves, one approach is to reduce the substrate thickness to about one tenth of the dielectric wavelength. This has a severe impact on the radiation efficiency of a microstrip antenna. One solution is using micromachining techniques, in which the substrate below the antenna element is removed, and in doing so synthesizing a region around the antenna with a low dielectric constant.

This technique was first investigated by Gauthier et al. [48] in the late 1990s; a definitive improvement on radiation efficiency was demonstrated at microwave frequencies (12.5–13.2 GHz). The authors declared that their technique should be readily applicable to millimeter-wave frequencies, and Papapolymerou, Drayton and Katehi published their findings at around 20 GHz [49] roughly a year later. Soon after, Gauthier et al. [50] presented a 94 GHz aperture-coupled microstrip antenna, which was fabricated using these micromachining techniques.

A cavity was etched into the substrate material below the antenna and the resulting synthesized dielectric constant was around 2.8–3.9, depending on the depth of the cavity. This dielectric constant was then used to determine the dimensions of the patch to enable it to radiate at 94 GHz. Compared to a rectangular slot, an H-shaped slot provides improved coupling and results in a higher front-to-back ratio.

To extend this element into an array configuration, a feed network based on coplanar waveguides was designed and a coplanar waveguide-to-microstrip transition was designed. Two variations were built around this transition, and the length of the input coplanar feed line was altered between these two designs. On a 100-μm thick silicon substrate, the radiation efficiency improved from 27 to 58 % when using a longer coplanar feed line. This variation also exhibited a large improvement in impedance bandwidth (for $S_{11} < -10$ dB) of 10 % (up from 3 %).

Generally, micromachined antennas can be divided into two groups. One approach uses a thin membrane to support the antenna, while the other uses an aperture-coupled antenna. Dielectric losses can be greatly reduced with the thin membrane approach and radiation efficiency can be improved, seeing that the dielectric constant underneath the patch is similar to that of air. On the other hand, sophisticated equipment is required to etch the substrate, since fabrication tolerances have a substantial effect at millimeter wavelengths. Furthermore, because the patch element presents a relatively high input impedance, a matching network is required, further adding to the complexity of the antenna, even more so when the element is extended into an array.

Implementing the feed network and radiating element on two different substrates, as is the case in an aperture-coupled antenna, requires precise alignment of the two substrates. Even minute cases of misalignment can result in a large degradation in performance.

In an attempt to solve the challenges posed by the two approaches discussed here, Kim et al. [51] developed a post-supported patch antenna, fed with a coplanar waveguide. The radiating patch is elevated from the substrate by two support pillars and one feeding pillar, creating an air gap between the feed network and the radiating element. The coplanar lines were designed in such a manner that matching sections were not required. In characterizing the performance of the antenna, the measured and simulated results were in good agreement. For the single element setup, a −10 dB bandwidth of 5.8 GHz was measured, stretching from 58.7 GHz to 64.5 GHz, equating to an impedance bandwidth of just under 10 %. The array configuration improved on this, resulting in a −10 dB bandwidth of 8.7 GHz, or 14.5 %.

Other notable patch antennas using this technique are the K_a-band cavity-backed patch antenna by Lukic and Filipovic [52] and a simple microstrip line-fed patch for 60 GHz applications, proposed by López et al. [53]. The antenna built by Lukic and Filipovic has a measured bandwidth of about 4 % around 36 GHz, with a radiation efficiency of over 95 % and a 6 dBi gain. On the other hand, the antenna built by López et al. has a measured impedance bandwidth of 5 % around 60 GHz, with a gain of 4.6 dBi.

3.4 Closing Thoughts

Throughout this chapter, we have presented a thorough investigation of printed and planar antennas. Emphasis has been placed on the fabrication difficulties related to millimeter-wave systems. As fabrication processes become more sophisticated, new designs become possible as is evident from micromachining antenna elements. The substrate-integrated waveguide has clearly played a substantial role in designing low-loss feeding structures, resolving much of the issues plaguing conventional feeding structures. Printed antennas tailored for millimeter-wave applications are evolving rapidly, particularly as a result of the mainstream development of new wireless standards and systems.

References

1. H. Gutton, G. Baissinot, Flat aerial for ultra high frequencies, French Patent 703113, 1955
2. P. Agrawal, M. Bailey, An analysis technique for microstrip antennas. IEEE Trans. Antennas Propag. 25(6), 4–7 (1977)
3. Y.T. Lo, D. Solomon, W. Richards, Theory and experiment on microstrip antennas. IEEE Trans. Antennas Propag. 27(2), 137–145 (1979)
4. W. Richards, Y. Lo, An improved theory for microstrip antennas and applications. Antennas Propag. Soc. Int. Symp. 17(1), 1979 (1979)
5. D.M. Pozar, Microstrip antennas. Proc. IEEE 80(1), 79–91 (1992)
6. D. Sengupta, The transmission line model for rectangular patch antennas. Antennas and Propag. Soc. Int. Symp. 21, 158–161 (1983)
7. C.A. Balanis, Antenna Theory: Analysis and Design, 3rd edn. (John Wiley & Sons Inc, Hoboken, New Jersey, 2005)
8. D.M. Pozar, Transmission Lines and Waveguides, Microwave Engineering, 4th edn. (John Wiley & Sons Inc, Hoboken, New Jersey, 2012)
9. K.C. Gupta, R. Garg, I.J. Bahl, Microstrip Lines I, Microstrip Lines and Slotlines (Artech House Inc, Dedham, Massachussets, 1979)
10. K. Carver, J. Mink, Microstrip antenna technology. IEEE Trans. Antennas Propag. 29(1), 2–24 (1981)
11. C.A. Balanis, Microstrip Antennas, Antenna Theory: Analysis and Design (John Wiley & Sons Inc, Hoboken, New Jersey, 2005)
12. M. Weiss, Microstrip antennas for millimeter waves. IEEE Trans. Antennas Propag. 29(1), 171–174 (1981)
13. F.K. Schwering, Millimeter wave antennas. Proc. IEEE 80(1), 92–102 (1992)
14. D.M. Pozar, Considerations for millimeter wave printed antennas. IEEE Trans. Antennas Propag. AP-31(5), 740–747 (1983)
15. N.G. Alexopoulos, P.B. Katehi, D.B. Rutledge, Substrate optimization for integrated circuit antennas. Microw. Symp. Dig. 1982 IEEE MTT-S Int. M(7), 550–557 (1982)
16. P. Katehi, N. Alexopoulos, On the modeling of electromagnetically coupled microstrip antennas-The printed strip dipole. IEEE Trans. Antennas Propag. 32(11), 1179–1186 (1984)
17. K. Wong, C. Tang, J. Chiou, Broad-band probe-fed patch antenna with a w-shaped ground plane. IEEE Trans. Antennas Propag. 50(6), 827–831 (2002)
18. A. Petosa, A. Ittipiboon, N. Gangon, Suppression of unwanted probe radiation in wideband probe-fed microstrip patches. Electron. Lett. 35(5), 355–357 (1999)

19. H.W. Lai, K.-M.L.K.-M. Luk, Design and study of wide-band patch antenna fed by meandering probe. IEEE Trans. Antennas Propag. **54**(2), 564–571 (2006)
20. T. Huynh, K.-F. Lee, Single-layer single-patch wideband microstrip antenna. Electron. Lett. **31**(16), 1310 (1995)
21. K.F. Tong, K.M. Luk, K.F. Lee, Design of a broadband U-slot patch antenna on a microwave substrate. Proc. 1997 Asia-Pacific Microw. Conf. **1**, (1997)
22. H. Sun, Y.-X. Guo, Z. Wang, 60-GHz circularly polarized U-Slot patch antenna array on LTCC. 2IEEE Trans. Antennas Propag. **61**(1), 430–435 (2013)
23. T. Baykas, C.S. Sum, Z. Lan, J. Wang, M.A. Rahman, H. Harada, S. Kato, IEEE 802.15.3c: The first IEEE wireless standard for data rates over 1 Gb/s. IEEE Commun. Mag. **49**(7), 114–121 (2011)
24. C.L. Mak, K.M. Luk, K.F. Lee, Geometry of wideband small-sized antenna: vertical patch antenna. Electron. Lett. **39**(25), 1777–1779 (2006)
25. K.L. Lau, K.M. Luk, K.F. Lee, A wideband C-shaped vertical patch antenna. Asia-Pacific Microw. Conf. Proceedings, APMC, **3**, 2024–2026 (2006)
26. K.C. Chao, F.S. Chang, H.T. Chen, C.H. Lu, Y.T. Liu, Dual-band operation vertical patch antenna for WLAN applications. IEEE Reg. 10 Annu. Int. Conf. Proceedings/TENCON, 3–5 (2007)
27. Z.H. Wu, E.K.N. Yung, Wideband circularly polarized vertical patch antenna. IEEE Trans. Antennas Propag. **56**(11), 3420–3425 (2008)
28. H. Wong, K.B. Ng, K.M. Luk, C.H. Chan, Q. Xue, Printed millimeter wave vertical patch antenna. 2010 Int. Conf. Commun. Circuits Syst. ICCCAS 2010—Proc., 647–649 (2010)
29. K.M. Luk, H. Wong, A new wideband unidirectional antenna element. Int. J. Microw. Opt. Technol. **1**(1), 35–44 (2006)
30. K. Luk, H. Wong, Complementary wideband antenna, US Patent 7,843,389, 2010
31. B.Q. Wu, K. Luk, A Broadband Dual-Polarized Magneto-Electric, vol 8, pp. 60–63, 2009
32. K.M. Mak, K.M. Luk, A circularly polarized antenna with wide axial ratio beamwidth. IEEE Trans. Antennas Propag., **57**(10), Part 2, pp. 3309–3312 (2009)
33. Z.Y. Zhang, G. Fu, S.L. Zuo, S.X. Gong, Wideband unidirectional patch antenna with r-shaped strip feed. Electron. Lett. **46**(17), 1238 (2010)
34. K.B. Ng, H. Wong, K.K. So, C.H. Chan, K.M. Luk, 60 GHz Plated through hole printed magneto-electric dipole antenna. IEEE Trans. Antennas Propag. **60**(7), 3129–3136 (2012)
35. K.B. Ng, C.H. Chan, H. Zhang, G. Zeng, Bandwidth enhancement of planar slot antenna using complementary source technique for millimeter-wave applications. IEEE Trans. Antennas Propag. **62**(9), 4452–4458 (2014)
36. M. Li, K.-M. Luk, A wideband circularly polarized antenna for microwave and millimeter-wave applications. IEEE Trans. Antennas Propag. **62**(4), 1872–1879 (2014)
37. H. Yagi, Beam transmission of ultra short waves. Proc. Inst. Radio Eng. **16**(6), 715–740 (1928)
38. J. Huang, Microstrip yagi array antenna for mobile satellite vehicle application. IEEE Trans. Antennas Propag. **7**(39), 1024–1030 (1991)
39. K. Uehara, K. Miyashita, K.-I. Natsume, K. Hatekeyama, Lens-coupled imaging arrays for the millimeter- and submillimeter-wave regions. IEEE Trans. Microw. Theory Tech. **40**(5), 806–811 (1992)
40. N. Kaneda, W.R. Deal, Y. Qian, R. Waterhouse, T. Itoh, A broadband planar quasi-yagi antenna. IEEE Trans. Antennas Propag. **50**(8), 1158–1160 (2002)
41. R.A. Alhalabi, G.M. Rebeiz, Differentially-fed millimeter-wave yagi-uda antennas with folded dipole feed. IEEE Trans. Antennas Propag., **58**(3), 966–969 (2010)
42. P.-Y. Qin, A.R. Weily, Y.J. Guo, C.-H. Liang, Millimeter wave frequency reconfigurable quasi-yagi antenna. 2010 Asia-Pacific Microw. Conf., 642–645 (2010)
43. C. Luxey, L. Dussopt, J.-L. Le Sonn, J.-M. Laheurte, Dual-frequency operation of CPW-fed antenna controlled by pin diodes. Electron. Lett. **36**(1), 2 (2000)
44. Z. Briqech, A. Sebak, Low-cost 60 GHz printed Yagi antenna array. IEEE Antennas Propag. Soc. AP-S Int. Symp. **1**, 7–8 (2012)

45. O. Kramer, T. Djerafi, K. Wu, Very small footprint 60 GHz stacked Yagi antenna array. IEEE Trans. Antennas Propag. **59**(9), 3204–3210 (2011)
46. O. Kramer, T. Djerafi, K. Wu, Vertically multilayer-stacked yagi antenna with single and dual polarizations. IEEE Trans. Antennas Propag. **58**(4), 1022–1030 (2010)
47. M. Li, K.M. Luk, Low-cost wideband microstrip antenna array for 60-GHz applications. IEEE Trans. Antennas Propag. **62**(6), 3012–3018 (2014)
48. G.P. Gauthier, A. Courtay, G.M. Rebeiz, Microstrip antennas on synthesized low dielectric-constant substrates. IEEE Trans. Antennas Propag. **45**(8), 1310–1314 (1997)
49. L. Papapolymerou, R.F. Drayton, L.P.B. Katehi, Micromachined patch antennas. IEEE Trans. Antennas Propag. **46**(2), 275–283 (1998)
50. G. Gauthier, J. Raskin, L. Katehi, G. Rebeiz, A 94-GHz aperture-coupled micromachined microstrip antenna. IEEE Trans. Antennas Propag. **47**(12), 1761–1766 (1999)
51. J.G. Kim, H.S. Lee, H.S. Lee, J.B. Yoon, S. Hong, 60-GHz CPW-fed post-supported patch antenna using micromachining technology. IEEE Microw. Wirel. Components Lett. **15**(10), 635–637 (2005)
52. M.V. Lukic, D.S. Filipovic, Surface-micromachined dual ka-band cavity backed patch antenna. IEEE Trans. Antennas Propag. **55**(7), 2107–2110 (2007)
53. A.V. López, J. Papapolymerou, A. Akiba, K. Ikeda, S. Mitarai, 60 GHz micromachined patch antenna for wireless applications, 2011, pp. 515–518

Chapter 4
Active Integrated Antennas

As miniaturization and the requirement for system-on-chip designs evolve, antennas that suit these applications are required, and recent years have seen a growing interest in such antennas. From the viewpoint of a microwave engineer, an integrated antenna may be considered an active system where the output or input ports are not conventional 50 Ω connections, but rather exist in free space. On the contrary, the antenna engineer may regard the integrated antenna as one in which filtering, mixing, and amplification are built into the antenna itself. Therefore, we can define the integrated antenna as one in which devices such as filters, phase shifters, mixers, and oscillators are integrated onto the same substrate as the radiating structure. The benefits of such an antenna are apparent; its reliability and compact nature make it a very attractive option for several applications. Research efforts from various groups have thus far established that these antennas should be useful for both the lower millimeter-wave region, as well as the upper millimeter-wave region above 100 GHz. The reasonably small size of the radiators, components, and circuit elements in the millimeter-wavelength regions is an excellent motivator for the use of monolithic and integrated systems.

4.1 Introduction

As high-frequency transistors became more prevalent in the electronics industry, a surge of interest in integrated antennas followed and several pioneering works were published in the 1960s and 1970s. Copeland, Robertson, and Verstraete experimented with the idea in the early 1960s, and one of their early reports detailed a dipole antenna integrated with a transistor amplifier, dubbed an "Antennafier"—an integrated antenna–amplifier configuration [1]. One of their publications described

© Springer International Publishing Switzerland 2016   61
J. du Preez and S. Sinha, *Millimeter-Wave Antennas: Configurations and Applications*, Signals and Communication Technology,
DOI 10.1007/978-3-319-35068-4_4

a broadside array consisting of four of these elements. It described a wide variation of beam control made possible by using different aperture distributions. However, the concept of integrated antennas was being explored much earlier, as early as 1928, when an antenna together with a wideband coupling tube was typically used in broadcasting [2].

Theoretically, low-frequency devices and antennas can be scaled up to millimeter wavelengths, resulting in high-frequency systems that can simply be developed from a low-frequency platform. However, factors in the design and fabrication processes can complicate matters. For example, microstrip lines and coplanar waveguides are very popular at microwave frequencies, but there are several disadvantages to millimeter-wave implementation [3]. Open guiding structures typically exhibit high dielectric losses (reducing the attainable Q-factor) and large current singularities over edges of the transmission line. Packaging issues begin to play a larger role, and crosstalk and DC grounding lead to cost issues. With the increase in frequency, power handling also becomes increasingly important and thermal management becomes a tedious issue, since operating TEM modes require separate conductors. Finally, impedance stability over wide bandwidths is difficult when extremely tight fabrication tolerances are required, necessitating advanced processing techniques.

As opposed to microstrip and coplanar lines, nonplanar transmission lines such as rectangular waveguides and coaxial guides are expensive and bulky, but nonetheless have been used in millimeter-wave components. These have the advantage of exhibiting low transmission loss with much higher power handling capability, as well as greatly improved electromagnetic shielding. Dielectric waveguides are another class of guiding structure, one that has not really been explored for millimeter-wave operation. This type of guide supports several dispersive modes, which are very difficult to separate. Furthermore, implementing an electrical interface with these guides is difficult, given that current generation active devices are specified with voltages and currents for TEM modes. Regardless of these difficulties, the low loss and reasonable cost make dielectric guides potentially attractive for millimeter-wave applications.

The noise figure of an RF system is heavily influenced by the components closest to the antenna [4]. Issues that relate to noise performance are accentuated above microwave frequencies and the effects on efficiency, impedance matching and loss become more apparent. By using a high-gain antenna, the requirements on the remainder of the front end are somewhat relaxed, and the implementation of a bandpass filter with an exceedingly sharp cutoff, as is quite common in point-to-point systems, could possibly be avoided.

The solution to the issues described here is an antenna that is completely integrated with its supporting circuitry on the same platform, using the same fabrication technique. Moreover, the development of the IEEE802.15.3c standard for 60 GHz communication networks has created a definitive requirement for compact on-chip, low-power 60 GHz transceivers.

4.2 Integrated Waveguides

In earlier chapters, we have briefly touched on substrate-integrated waveguides as a new type of guiding structure that is suitable for millimeter-wave circuits. Given our objective for this chapter, a more detailed (albeit brief) look at the various types of integrated waveguides is necessary.

The concept of substrate-integrated circuits eventually led to the implementation of the technology for use in antennas. The technology essentially enables conversion of nonplanar structures into a planar form, thereby enabling planar processing techniques [5]. The result is that conventional planar transmission lines can be integrated together with the nonplanar structures, on the same substrate and through the same fabrication process. The three most widely used techniques are the substrate-integrated waveguide, the substrate-integrated non-radiative dielectric guide, and the substrate-integrated image guide.

4.2.1 Substrate-Integrated Waveguide

The substrate-integrated waveguide is synthesized with two rows of plated vias (or slot trenches) that are embedded in a substrate. The top and bottom broadside walls are then metalized as well, creating an integrated structure similar to a rectangular waveguide. As a result of the dielectric filling, the spacing between the two rows of vias determines the cutoff frequency. Since there is essentially no current flow along the two side walls, this guide can only support TE_{n0} modes. This type of guide retains most of the advantages offered by the conventional rectangular waveguide, such as a high Q-factor and excellent power handling capability, as well as consistent mechanical and electrical shielding properties. An illustration of this waveguide is shown in Fig. 4.1.

Several experimental and theoretical studies have been conducted on substrate-integrated waveguides in the last decade, typically with the focus on implementing such networks with various fabrication techniques. In recent years, many different topological variants of the guide shown in Fig. 4.1 have been proposed and successfully implemented. Some of these are shown here in Fig. 4.2.

Most of these alternate topologies are driven by the requirement to improve the compactness of the structure. The half-mode substrate-integrated waveguide, for example, permits size reduction of nearly 50 %, which is quite significant. Building on this approach, the folded half-mode guide represents an attempt at further size reduction [7]. An early solution to enhancing the operating bandwidth came in the form of the slab guide, where the original substrate-integrated waveguide structure is periodically perforated with holes in the dielectric region. Furthermore, introducing a ridge into this structure in the form of thin metal posts located in the center

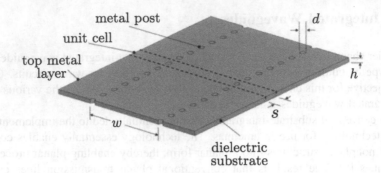

Fig. 4.1 Geometry of a substrate-integrated waveguide. Reproduced by permission of the Institution of Engineering and Technology [6]

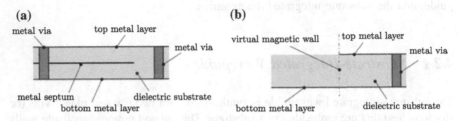

Fig. 4.2 Topologies of substrate-integrated waveguides. **a** Folded waveguide. **b** Half-mode guide. Reproduced by permission of the Institution of Engineering and Technology [6]

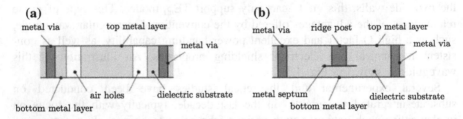

Fig. 4.3 Bandwidth enhancement topologies of the substrate-integrated waveguide; **a** slab guide and **b** ridge guide. Reproduced by permission of the Institution of Engineering and Technology [6]

of the waveguide leads to a bandwidth enhancement of 73 % (up from 40 % in the case of the slab guide) [8]. By modifying the ridge waveguide to use partial height metal cylinders in the larger waveguide wall, a bandwidth enhancement of 168 % is achievable [9]. These topologies are shown in Fig. 4.3.

4.2.2   Substrate-Integrated Non-radiative Dielectric Guide

Non-radiative dielectric guides originally consisted of dielectric slabs placed between two metallic plates. The separation distance between the plates determines the frequency range, while the slab width controls the bandwidth. The substrate-integrated version of this guide uses a material with a lower dielectric constant than the center channel to cover the bilateral extent of the structure. Thus, two dielectric regions are created.

In the early 1980s, Yoneyama et al. [10] reported on the use of the non-radiative dielectric guide for millimeter-wave integrated circuits, publishing results of up to 50 GHz. An important advantage of this guiding structure is that it is already easily integrated with planar circuits as well as hybrid planar techniques [11], but it does suffer from lack of mechanical support and the assembly of a planar substrate can prove to be difficult. Cassivi and Wu [12] were among the first researchers to use a specific pattern of air-filled holes to lower the dielectric constant of a particular region within the substrate, effectively creating a dielectric waveguide in the substrate material. A complete design guideline was developed along with this novel fabrication technique and the prototype waveguide (measured from 34 to 38 GHz) exhibited excellent correlation between theoretical predictions and practical measurements.

4.2.3   Substrate-Integrated Image Guide

Another project in which Wu was involved was, this time teamed up with Patrovsky, synthesizing a conventional dielectric image guide into a substrate-integrated configuration [13, 14]. An illustration of the conventional insulated image guide is shown in Fig. 4.4.

In order to create dielectric regions with lower permittivity values, a laser-cut triangular lattice of air-filled holes was inserted into low-loss, high-permittivity substrates. This technique is suitable for low-cost integrated circuits and enables repeatable, high-precision fabrication.

Fig. 4.4 Insulated image guide

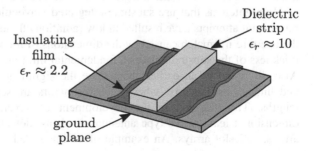

Insulating film $\epsilon_r \approx 2.2$

Dielectric strip $\epsilon_r \approx 10$

ground plane

In its most rudimentary form, the dielectric image guide consists of a strip of dielectric material placed on top of a ground plane. This ground plane is effectively a mirror plane for the fields that exist in the strip. The conductive ground can be used to bias or dissipate heat from active devices and is the only conductor involved in the absence of any field singularity. Thus, conductor losses are reduced as a result of lower current densities. The dimensions of image guides are in the order of half a wavelength, which is significantly larger than other printed circuit guides such as microstrip lines. This could be considered a disadvantage in terms of miniaturization, but in fact could be beneficial for low-cost production because the mechanical tolerances are less stringent.

4.3 Classification of Integrated Antennas

Our discussions on leaky-wave antennas and printed antennas in the previous chapters referred to several of these highly versatile integrated antennas. This chapter will therefore be subdivided into applications of integrated antennas within RF systems, as opposed to our previous approach of grouping sections in the text according to physical structure or principles of operation. At the top level of sub-division, integrated antennas can be separated into two functional categories, namely transmit and receive antennas. The transmit and receive chains in an RF network vary in terms of their accompanying circuitry, for example a detector circuit will be exclusive to the receiving antenna, and so forth. Furthermore, antennas that do not necessarily fit either of these roles (or perhaps both) will be discussed separately. These may include integrated phased arrays and beamforming networks. As one could expect, many groups have developed phased array systems that only consist of a receiver or transmitter module and these papers could be covered in all subsections of this chapter.

Nonetheless, some architectures are encountered quite often when millimeter-wave phased array systems are discussed, the first of these being slot arrays. Discussed in an earlier chapter, these antennas consist of a waveguide, where periodic perturbations of different shapes and sizes are cut into either the broad or narrowwalls of the structure. As a result, power leakage occurs outward from the waveguide walls and this technique is equally applicable to conventional rectangular waveguides as to the integrated waveguides discussed in this chapter.

Horn antennas that use substrate-integrated waveguides are another possibility, but several attempts have resulted in low radiation efficiency, primarily as a result of the aperture mismatch caused by the dimensions of the waveguide. In this case, the thickness of the substrate places a fundamental limit on the dimensions of the horn. Also discussed in an earlier chapter are the significant manufacturing difficulties and limitations that arise when patch resonators are scaled into millimeter wavelengths. This has led to the development of several surface-wave and two-dimensional leaky-wave type antennas, such as dielectric rods, printed Yagi-Uda arrays, and slot arrays. An example of an integrated dipole is shown in Fig. 4.5.

Fig. 4.5 A 77 GHz on-chip dipole antenna, designed for automotive radar. Image reprinted with permission of Sabanci University, Istanbul

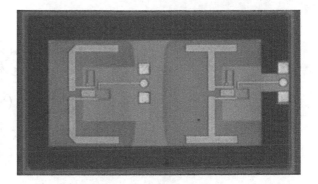

The antenna consists of two on-chip dipoles, one L-shaped, and one T-shaped, integrated with a lumped LC balun circuit. The circuit is fabricated using a 0.25 μm silicon-germanium bipolar complementary metal oxide semiconductor (SiGe BiCMOS) process. With the proposed balun circuit, an impedance bandwidth of up to 12 GHz has been achieved [15].

Dielectric resonators have recently been suggested as a potential alternative for millimeter-wave systems and some configurations consist of a dielectric resonator fed by a half-mode substrate-integrated waveguide [16, 17].

The configurations mentioned here represent passive integrated antennas and for the purposes of this discussion, it is important to note the difference between passive and active integrated antennas. Active antennas refer specifically to the circuit-antenna modules that we have touched on thus far, where ancillary circuitry is fabricated onto the same substrate as the antenna itself. See for example the integrated patch antenna shown in Fig. 4.6. The antenna is implemented on an electromagnetic bandgap (EBG) substrate, which serves to suppress spurious radiation from surface waves. The substrate below the patch is removed, and replaced with another substrate with a lower dielectric constant [18].

Fig. 4.6 An integrated 77 GHz microstrip patch antenna. Image reprinted with permission of Sabanci University, Istanbul

Fig. 4.7 Block diagram of a generic quadrature transceiver system

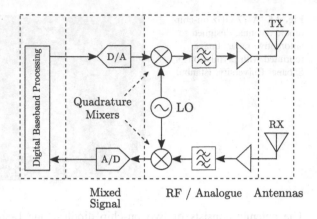

Mixed   RF / Analogue   Antennas
Signal

Furthermore, this class of antenna includes those that can be reconfigured using an on-board active device. Seeing that passive integrated antennas often overlap with other chapters of this text, this chapter will focus on active devices and several of the configurations mentioned here will be discussed further in succeeding sections. To aid in understanding the concept of an integrated antenna, consider the block diagram shown in Fig. 4.7.

From this diagram, we can roughly establish the composition of an integrated antenna system. Mixers, filters, and amplifiers (typically low-noise amplifiers for the receiver chain, and power amplifiers for the transmitter chain) are perhaps the simplest components that are integrated with the antenna on a suitable substrate.

4.4   Integrated Transmitter Antennas

Integrated transmitter antennas are those that integrate elements of the transmit chain (such as the simple illustration in Fig. 4.11) onto the antenna substrate. An example of an on-chip transmitter is shown in Fig. 4.8.

Camilleri and Bayraktaroglu [19] were some of the earliest researchers to develop such antennas successfully and one of their works, published in the late 1980s, demonstrated a loop antenna operating at 43.3 GHz with a set of on-chip IMPATT diodes. The diode resonator acted as a radiating element, meaning that the oscillator was inherently matched to the antenna, thereby eliminating mismatch losses in the feed system. Experimental results yielded 27 mW of output power while operating in CW mode, with a 7.2 % conversion efficiency. The dimensions of the prototype chip were 1.25 mm × 0.75 mm. Thoren and Virostko [20] also developed a solid-state transmitter with IMPATT diodes, using them in a power combiner network. The system was capable of delivering 1.89 W peak power, and designed to have tunable pulse width from 0.1 μs up to 4 μs and tunable duty cycles between 5 and 35 %. The output frequency was mechanically tunable from 90 to 99 GHz, with a maximum of 1 GHz injection gain.

Fig. 4.8 60 GHz 2 × 2
phased array transmitter.
Image reprinted with
permission of RICE
University

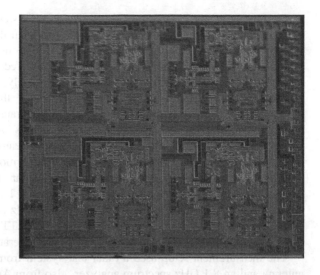

Exploring the same concept, Itoh and Birkeland [21] demonstrated a slightly different implementation of an integrated transmit antenna. The radiating element was linear microstrip patch array fed in series, while a negative resistance field-effect transistor (FET) was used as the oscillator. Alternatively, periodic dielectric radiators using integrated Gunn diodes as the oscillator elements were studied by Itoh and Song [22].

A different approach to integrated transmit antennas uses quasi-optical power combiners, like the ones investigated by Mink [23]. These combiners make use of an open resonator to injection lock an array of integrated solid-state sources, which are distributed over the surface of a reflector. An excited resonant mode with a high Q-factor forces all the sources to operate in frequency and phase coherence. A second reflector, which is partially transparent, is used to extract the combined power, in the form of a coherent wave. One key advantage of these types of combiners is that their dimensions can span many wavelengths. The only remaining requirement to use these combiners at millimeter wavelengths is implementing an appropriate modulation technique for the sources.

Shaping of the transmit beam is achieved by choosing suitable geometry for the reflectors, although several authors have illustrated that the relationship between the reflector geometry and the resulting beam shape is quite complex. The resonator unit reported by Mink consists of a planar and curved reflector (with a uniform curvature profile) to produce a narrow Gaussian main beam with very low side-lobes, which are ideal attributes for a multitude of applications.

Quasi-optical combiners were first investigated and reported on by the Rutledge group [24], but also by Compton and York [25] early in the 1990s. A very admirable quality of the Fabry–Perot resonators is their simple geometry, making their extension into millimeter wavelengths relatively simple.

Designed for millimeter-wave automotive radar systems around 77 GHz, Stiller et al. developed a monolithic integrated transmitter in the mid-1990s, which consists of a slotted patch resonator and an IMPATT diode [26]. The microstrip resonator is 2.97 × 1.97 mm in size, and it is center-fed from a 1.6 mm long slot resonator. The slot length corresponds to approximately one wavelength at 75 GHz and the IMPATT diode is placed in the center of the slot. The configuration described here is fabricated in a 100 μm silicon substrate and it is advantageous for a number of reasons. The impedance specifications of the IMPATT diode are satisfied, even on a thin grounded substrate, which means that a heatsink can easily be added to the ground layer of the substrate. Furthermore, no vias are required to integrate the IMPATT diode with the coplanar slot resonator structure. Measurements of the experimental system revealed a 1 mW maximum CW power level at 79.7 GHz, with a phase noise of 81.7 dBc/Hz at a 100 kHz offset.

Al-Attar and Lee developed an integrated IMPATT transmitter, fabricated in standard 0.18 μm CMOS technology for 77 GHz operation [27].

The measurement setup used a radar test system from Anritsu, an E-band horn antenna and an 8.1 GHz spectrum analyzer, also from Anritsu. The received signal is down-converted by the radar test unit to an intermediate frequency (IF) of 4.7–5.7 GHz, and thereafter the signal is characterized by the spectrum analyzer. With the biasing voltage of the on-chip transmitter at 11 V, the quiescent current is approximately 30 mA. The transmitter delivers up to −62 dBm of output power, and the resulting diode efficiency has been calculated as −41 dB. This is fairly low efficiency, primarily because of a high capacitive loading that occurs from the depletion region.

Dawn et al. [28] demonstrated two 60 GHz transmitters, both developed in a 90 nm CMOS process. One setup focused on low-power applications, while the other was aimed at high performance systems. The former transmitter includes a push-push voltage controlled oscillator (VCO), a three-stage power amplifier and a single-gate mixer, while the latter consists of a cross-coupled VCO, a three-stage power amplifier, and a double-balanced Gilbert cell mixer with an on-chip Marchand balun. Wide tuning ranges (in excess of 2 GHz) were achieved for both VCO designs. The single-gate mixer was found to consume 28 mW of power, while providing an 8.4 dB conversion loss, compared to the 4 dB conversion loss for the Gilbert cell mixer.

The low-power transmitter provides an 8.4 dB gain with 5.7 dBm saturated power output, while the high performance version provides a 12 dB gain with a saturated power output of 8.6 dBm. Both modules achieve a 3 dB bandwidth greater than 8 GHz, between 57 and 65 GHz.

Shin et al. [29] recently developed a wafer-scale integrated phased array transmitter for operation in the 108–114 GHz band. The 16-element array is fabricated on a 6.5 mm by 6.0 mm chip using an SBC18H3 SiGe BiCMOS process, and the antennas are implemented on a 100 μm quartz substrate. From a 1.9 V power supply, the system consumes 3.4 W during its normal operation. The prototype system achieved a 26.5 dB array gain and a directivity of 17 dB at 110 GHz.

4.5   Integrated Receiver Antennas

These devices combine the function of a receiving antenna with a detector and mixer in one package. Typically, the output terminals of the antenna and the input terminals of the mixer section are spatially close to one another, shortening the RF path (and in some cases, completely eliminating it) between the two blocks and potentially reducing the overall noise figure. Coupling the oscillator signal with the mixer may be achieved by using a waveguide, but it is also possible to receive it in a quasi-optical fashion through the antenna. Itoh et al. [30] demonstrated a simple receiver antenna with an integrated diode-based mixer. The system was designed for operation at 65 GHz and the integrated mixer exhibited a 6.5 dB conversion loss in experimental measurements. Their antenna consisted of a circularly shaped slot, etched into a metal cladding that covered one side of a substrate layer. The balanced mixer diodes were inserted at two positions, each offset from the vertical axis by ±45°. The LO signal then passed through the diodes, down-converting the RF signal at each diode. The two IF signals were then combined through a low-pass filter, which was subsequently coupled to a coplanar waveguide.

4.5.1   Monopulse Antennas

High-resolution tracking applications such as communications satellite tracking or perhaps antimissile terminal guidance typically employ millimeter-wave tracking radars. Monopulse tracking radars are extensively used because of their tracking reliability and inherent resistance to countermeasures [31]. These radars can operate in phase comparison or amplitude comparison modes and despite the fact that phase comparison is the most common approach, a combination of both methods may be employed as well [32, 33]. Common to any monopulse system is a microwave device that can produce sum and difference patterns from a set of directional antennas. To illustrate this, consider the simplified block diagram of a monopulse radar in Fig. 4.9 [31].

Fig. 4.9 Simplified block diagram of a single-channel monopulse radar

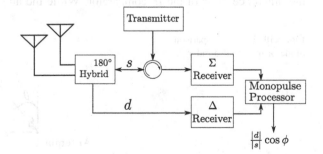

In the transmission mode, the sum port on the 180° hybrid is excited and the two antennas are excited with in-phase signals of equal amplitude. Upon reception, the hybrid splits the signals from the two antennas into a sum signal, which is used to normalize the monopulse ratio, and a difference signal, which is used to estimate the direction of the target. In the phase comparison mode, the phase difference is used to estimate the target direction (Φ in Fig. 4.9). The phase comparison concept is illustrated in Fig. 4.10 the comparison process facilitated by the hybrid coupler generally takes place at RF, as close to the antennas as possible, to reduce the effects of system noise and nonlinearity.

For these systems, an integrated approach can offer cheaper, more compact front-end modules than alternative, waveguide-based systems. Raman et al. [34] demonstrated an integrated monopulse receiver for W-band (94 GHz) radar. The receiver system is supported by a dielectric lens, with slot-ring antennas fed by coplanar waveguides, integrated with subharmonic mixers. As opposed to the common approach of RF comparison, this system forms the sum and difference signals at an IF between 2 and 4 GHz. This is due to the ability of IF processing to produce much deeper nulls in the difference patterns at millimeter wavelengths, since phase and amplitude tuning sections can be added before the monopulse network. In order to reduce (and in some cases, eliminate) substrate modes, a slot antenna is placed on a dielectric lens with more or less the same permittivity as the antenna wafer. The slot is fed by a coplanar waveguide. The lens appears as a dielectric half space, which inherently cannot support the excitation of surface waves.

Subharmonic mixing was utilized because it enables easy distribution of the LO signal at 45 GHz and provides good isolation between the RF and LO ports, which aids in limiting the leakage power from the LO to the antennas. Experimental measurements of the system at 91 GHz included losses that resulted from backside radiation (0.2 dB) and lens reflections (2.7 dB), feed line losses (1 dB), and mixer conversion loss (5 dB). The monopulse patterns were measured by illuminating the 24 mm lens aperture with a W-band signal, chopped at 1 kHz, and thereafter detecting and measuring the sum and difference outputs with a lock-in amplifier. The measured sum pattern was found to be rotationally symmetric, with a 25° Rayleigh (peak-to-null) beamwidth. Cross-polarization levels peaked at −25 dB. Sidelobe levels were in the order of −13 to −15 dB. Individual null depths were measured with a spectrum analyzer in order to determine the peak-to-null ratios at the difference ports of the IF comparator. While the amplitude matching was very

Fig. 4.10 Phase comparison mode of monopulse radar

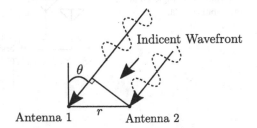

good between the four channels, significant phase adjustments were required to maximize the null depths on boresight. To improve the dynamic range, a DC −6 GHz LNA with a 3 dB noise figure and a gain of 35 dB was placed before the spectrum analyzer connection. Null depths of greater than 45 dB were obtained, corresponding to angular accuracy that is in the milliradian range.

Another monolithic monopulse transceiver was demonstrated by Niehenke et al. in the early 1990s, which was designed to operate at 94 GHz and provided switchable polarization modes [35]. Integrated circuits included mixers, IF amplifiers, and PIN diode switches. Although the microstrip patch antennas were not integrated on the same substrate, this design was an early example of the move toward on-chip radar that occurred near the turn of the century.

Collaboration between Lockheed Martin, the University of Massachusetts, the Syracuse Research Center and the Naval Research Center in the mid-90s yielded the first active phased array monopulse receiver, designed to operate at 94 GHz [36]. This system provided electronically scannable monopulse patterns with beamwidths of 2.3°, with null depths reaching up to 20 dB in both azimuth and elevation planes. A crucial component in the array construction was an eight-channel module, referred to as an 'octopak'. Each of these modules consisted of eight radiators, each integrated with its own LNA and phase shifter, thereafter followed by a microstrip combiner network. The narrow beamwidth was obtained by combining these octopak modules into two independent linear arrays, each consisting of 32 elements. The two arrays were then placed in an orthogonal configuration (i.e., to form a crossed-line array). The phase shifters used were monolithic GaAs varactor controlled, with the control words being supplied through an 8-bit digital-to-analogue converter. With this phase adjustment scheme, a worst case quantization error of 7.6° was observed.

The array beamforming was realized with a two-step waveguide combining approach. The first step consisted of combining the outputs of two octopaks through the sum port of an in-phase power combiner. In the second step, a magic-T network accepted the input signals from the aforementioned combiners and produced the sum and difference signals at its output. The diameter of the array was less than 75 mm.

4.6 Integrated Phased Arrays

Silicon technology creates many possibilities for microwave and millimeter-wave applications. The ultra-high cutoff frequencies that are possible with bipolar SiGe heterojunction transistors and the ever-decreasing feature dimensions of metal oxide semiconductor field-effect transistor (MOSFETs) require new design techniques to handle high-frequency coupling, lossy substrates, and parasitics that arise from interconnects. One of the first integrated phased arrays in silicon technology (operating at 24 GHz) was proposed by Hajimiri et al. [37]. The ability to integrate complete phased arrays in silicon leads to significant improvements in cost, reliability, reproducibility, and size, while simultaneously creating the possibility of on-chip signal conditioning and processing. When considering the complete system, this approach could lead to additional cost and power savings.

4.6.1　Phased Array Architecture

Generally, phased arrays are used to imitate highly directional antennas with electronically steerable beam directions, eliminating the requirement for cumbersome mechanical steering mechanisms. Moreover, the parallel nature of a phased array transceiver system reduces the noise figure and power handling specifications for individual active devices placed throughout the array. The conventional approach to phased arrays involves using large numbers of microwave integrated circuits, which greatly contribute to the overall system cost, seeing that many commercial phased array systems consist of several hundred antenna elements.

A phased array system consists of multiple signal paths. Each is connected to an antenna element, which in turn can be arranged in multiple spatial configurations [38]. When operated in the transmitter mode, each element is excited with a signal that is progressively phase-shifted (time-delayed) with respect to the previous one, to a point where signals add in phase in only one direction, ideally. A block diagram of the transmitter chain is shown in Fig. 4.11.

The vast majority of modern RF systems employ quadrature modulation schemes in order to retain phase and amplitude information in the signal. To implement the phase shifters shown in Fig. 4.11, a quadrature architecture is required. This is illustrated in Fig. 4.12.

Realizing the phase shifter blocks shown here can be achieved with multiple different approaches, as outlined by Hashemi et al. [39].

Similarly, in the receiver mode, a signal that is misaligned with the antenna phase front will arrive at different times at each element. This time difference is dependent on the angle of arrival (i.e., the angle between phase fronts), as well as spacing between individual elements. The receiver thus compensates for the individual phase shifts and adds the received signals coherently only when they

Fig. 4.11 Phased array transmitter

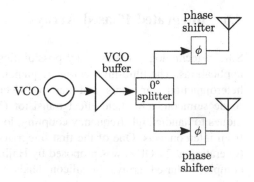

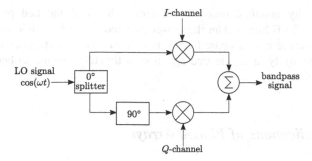

Fig. 4.12 Block diagram of a quadrature modulator. The individual phase of the I and Q-channel signals can be controlled to ultimately generate a phase-shifted bandpass signal

originate from the desired direction. To illustrate how the antenna creates a phase front that is offset from the main beam direction, consider the diagram in Fig. 4.13.

The phase shift of the excitation signal fed into each antenna element can be individually controlled, and depending on the particular phase-shifting profile used, a beam that is offset from the principal direction is created.

The result is that a phased array transmitter generates less interference with receivers that are anywhere but collocated with the main beam of the array. Furthermore, for a given specification on minimum detectable signal level at the receiver, a phased array requires less power to be generated when compared to an isotropic transmitter. Consider a transmitter with m total antenna elements, where each element is individually capable of radiating P watts. Therefore, the total power available at the receiver at the desired incidence angle is $m^2 P$ watts, where the squared value originates from coherently adding fields at the specified incidence

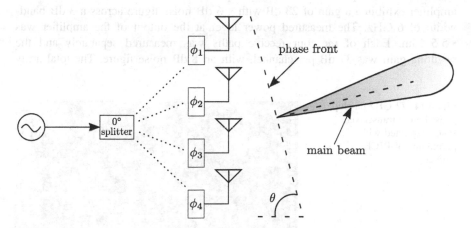

Fig. 4.13 Generating an offset phase front in a phased array antenna. The excitation to each element is progressively phase-shifted, Φ_4 indicating the largest phase shift

angle. For a hypothetical four-element array, the total radiated power can be computed as 12 dB higher than the power radiated by each individual element.

On the receiver end, a phased array is particularly advantageous owing to its improved sensitivity in the desired rejection, ultimately leading to better interference rejection.

4.6.2 Applications of Phased Arrays

Integrated phased arrays have received significant attention in the technical community, even more so as integrated circuit technology evolved in the past few decades. A 77 GHz phased array transceiver developed by the Integrated Circuits and Systems group at RICE University is shown in Fig. 4.14.

The system consists of four transmitters and four receivers, with a total of eight on-chip lens-coupled dipole antennas. Continuous phase rotators are used for electronic steering, and the circuit is fabricated with a 130 nm SiGe BiCMOS process.

Hajamiri et al. demonstrated a four-element phased array transceiver for automotive radar applications, operating in the 77 GHz band. The receiver [40] and transmitter [41] sections were detailed in two separate publications. The on-chip receiver that was developed included a complete down-conversion path, with low-noise amplifiers, phase shifters, a frequency synthesizer, combining amplifiers as well as integrated dipole antennas. To implement the signal-combining function, a novel distributed active combining amplifier was developed at a 26 GHz IF. Furthermore, the output of a 52 GHZ VCO was routed to various elements; each path could be individually shifted in phase. In order to reduce losses that result from the excitation of surface waves, a silicon lens was used on the back end of the module.

The measured gain of the on-chip dipoles was 2 dBi, while the low-noise amplifier exhibited a gain of 23 dB with a 6 dB noise figure across a 3 dB bandwidth of 6 GHz. The measured power level at the output of the amplifier was −5.5 dBm. Each of the four receive paths was measured separately and the resulting gain was 37 dB per channel, with an 8 dB noise figure. The total array

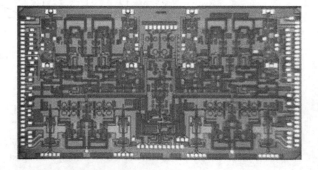

Fig. 4.14 77 GHz on-chip phased array transceiver. Image reprinted with permission of RICE University

gain was measured at 49 dB across a 3 dB bandwidth of 2 GHz. Compared to the configuration that did not include the silicon lens on the back end, the gain was improved by approximately 10 dB.

On the transmitter end, the module used phase-shifting in the local oscillator (LO) path to achieve beam steering. Every element in the four-element array generated 12.5 dBm of output power across a 2.5 GHz bandwidth (centered at 77 GHz), which equates to an isotropic effective radiated power of 24.5 dBm. Furthermore, each of the power amplifiers implemented on-chip had a saturation power level of 17.5 dBm at 77 GHz. With two transmit and two receive antennas active, the array achieved a 12 dB peak-to-null ratio, measured using an internal test setup.

Implementing the phase shifters in the LO path is advantageous, seeing that components in the LO path are operated in saturation, which means that it is easier to hold the gain of each element at a stable value (equal to the other elements). In other words, the gain of the element remains constant with various phase shift settings. Moreover, the required linearity, noise figure and bandwidth on the phase shifter section are significantly relaxed with this approach. Earlier demonstrations of integrated phased array transceiver systems used a centralized phase-shifting scheme implemented in the LO path as well, where a VCO generates several phase-shifted versions of the LO signal [42, 43]. Based on a desired pointing angle for the main beam, a phase selector then decides between the appropriate LO signals in order to generate the offset beam. However, a key limitation to this approach is that phase resolution is limited, since the VCO can only generate a finite number of phases and the complexity of the phase selector further limits the achievable resolution.

Another limitation, which is slightly more prevalent at millimeter-wave frequencies, is the fact that all the available LO signals need to be distributed to each antenna element. This requires a complicated distribution network, since all the LO signals need to be power-matched, intermediate buffers need to be matched and a transmission line interconnect is required between sections. Therefore, with larger arrays and at higher frequencies, the centralized approach is not desirable, since the LO distribution network would scale to exponentially larger sizes.

Natarajan et al. [44] recently demonstrated a 16-element phased array receiver, which is fully integrated using 0.12 μm SiGe BiCMOS technology. The system has been designed for multi-Gbps communication networks in the 60 GHz ISM band (57–66 GHz). The down-conversion architecture is a modified sliding-IF super-heterodyne approach. Such an architecture is shown in Fig. 4.15.

The phase-locked loop (PLL) generates LO signals at 58.32, 60.48, 62.64, and 64.8 GHz, as specified in IEEE 802.15.3c, from a 308.5714 MHz reference clock. An on-chip VCO with a tuning range of 16–18.6 GHz drives a frequency multiplier, and this LO signal is then used for the first down-conversion stage. These LO signals are centered at 49.99, 51.84, 53.69, and 55.54 GHz.

Part of the IF amplifier circuit is a programmable resonant load that holds the sliding-IF frequencies at their specified points, viz., 8.33, 8.64, 8.95, and 9.26 GHz. In order to implement phase correction to the quadrature LO signals, programmable

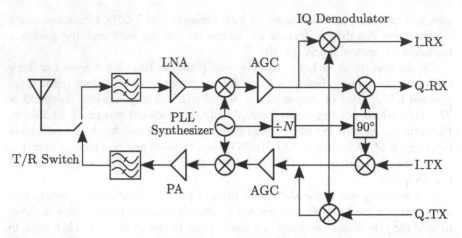

Fig. 4.15 Block diagram of a sliding-IF transceiver architecture [45]

phase rotators are used to drive the down-conversion mixers. The phase rotators are fed from the output of a factor of two frequency divider, which produces the quadrature LO signals in the first place. Channel filters at baseband were not included in the design; instead they were fabricated as standalone chips.

Several phase-shifting architectures have been used in millimeter-wave phased arrays. Kishimito et al. [46] used an IF/baseband approach, where a phase shifter block changes the phase of the RF signal by switching signal paths at baseband. We have discussed the LO path approach used by Guan et al. [43], but RF phase shifting remains the most common [44, 47, 48]. RF phase shifters and combiners require the lowest number of circuit blocks for each element in the phased array, at the cost of a possible degradation in performance that results from phase shifter linearity, noise, and loss. Power consumption, noise figure, and insertion loss (as a function of phase shift) must be considered to be able to choose the optimal approach for a given application. Natarajan et al. reported that with an LNA gain of 22 dB and a 6 dB noise figure, cascaded blocks having a 15 dB noise figure degrade the noise figure in the front end by less than 0.25 dB. Consequently, either active vector interpolators or passive reflection-type phase shifters are suitable options, but the lower power consumption of the passive option is an attractive characteristic for mobile systems.

However, these passive phase shifters exhibit losses that are dependent on the phase shift setting, because of the variable quality factor of the varactor. To compensate for this insertion loss, a variable-gain amplifier (VGA) can be used. This also serves to facilitate additional amplitude control to allow for pattern synthesis, but the on-chip implementation of these VGAs could prove to be quite difficult and adds additional power consumption into the system. This approach can be regarded as hybrid RF path passive-active phase shifting. The noise figure for each RF front end was observed at 22 and 65 °C, and the values were 6.8 and 7.5 dB, respectively. The receiver was capable of switching beams within 50 ns, and when used with a companion transmitter IC, line of sight data rates reached up to 5.3 GB/s using an orthogonal frequency division multiplexing (OFDM) scheme.

In a subsequent issue of the *IEEE Journal of Solid-State Circuits*, Okada et al. [49] published a report on their IEEE802.15.3c integrated transceiver module. The module uses a direct conversion approach using quadrature oscillators and is fabricated with a 65 nm CMOS process. Included in the package is a transmitter with an 18.3 dB conversion gain, 10.9 dBm output power saturation and 8.8 % efficiency, as well as a receiver with a 17.3 dB conversion gain and a noise figure below 8 dB. The frequency synthesizer is designed to produce a 60 GHz LO signal through a 20 GHz PLL and a 60 GHz quadrature, injection locked oscillator. The oscillator achieves a phase noise of −95 dBc/Hz at 60 GHz.

Shahramian et al. [50] demonstrated data rates of 10 Gbps (using 16-QAM, over a 1 m link) with a 70–100 GHz direct conversion phased array chipset. To fabricate the chips, a 0.18 μm SiGe BiCMOS process, containing four antenna elements and a set of direct conversion mixers, was used. The prototype receiver achieved a conversion gain of 33 dB with a noise figure of less than 7 dB between 75 and 95 GHz. Each of the four transmitter channels provides a flat output saturated power of at least 5 dBm. Both modules support low voltage operation between 1.5 and 2.5 V, and each consumes about 1 W during normal operation. The reported data rate is achieved at a frequency of 88 GHz and changing the modulation scheme to 32-QAM lowers the data rate reached with a 1 m link to 8.75 Gbps.

4.7 Integrated MIMO Antennas

As an alternative to the phased array architecture, multiple input, multiple output (MIMO) configurations have only recently been seriously investigated. A MIMO architecture (as shown in Fig. 4.16), in conjunction with an integrated transceiver, could provide several benefits for radar systems [51]. For one, using multiple transmitters means that a target of interest is illuminated from multiple directions, which can potentially negate the angle dependency of radar cross-section (RCS). However, the radar antennas need to be placed at a considerable distance from one another for this to truly have an effect [52]. Another benefit is the ability to synthesize multiple antenna elements, which is used to create virtual antenna positions, leading to a larger number of effective elements in the array.

Therefore, it is possible to increase the number of unambiguous targets in the same range bin, when compared to conventional radar architectures [53, 54]. A MIMO system can also be mapped to a conventional radar, and virtual receiver positions can be synthesized corresponding to the spatial convolution of the receiver and transmitter phase centers. This increases the number of virtual channels, which leads to a larger virtual array aperture and in turn provides improved angular resolution. A combination of phased arrays and MIMO architectures has been investigated for use in radar applications [55, 56].

Fig. 4.16 MIMO system
architecture

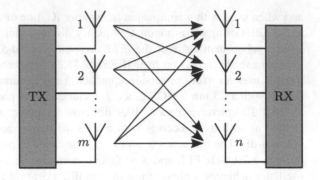

A MIMO frequency modulated continuous wave (FMCW) radar was developed
with a SiGe single-chip transceiver by Feger et al. [57]. A MIMO architecture is
adopted to improve angular resolution while limiting the chip size and number of
channels, and the prototype MIMO system allows synthesizing of ten different
element positions from just four physical channels. A further increase in angular
resolution was achieved by using a sparse array, which allows for an increase in
element spacing while avoiding the ambiguities that result from spatial under-
sampling. The increased spacing also serves to reduce mutual coupling and
therefore reduces the resulting loss. A rat-race coupler (like the one illustrated in
Fig. 4.17) is used with two switchable amplifier blocks to facilitate reconfiguring
the module between transmit and receive modes. A similar approach was followed
in an earlier publication [58]. Four of these cells were integrated with a 77 GHz
VCO, Wilkinson power dividers and a frequency divider, used to derive LO signals
from the VCO. A synchronous serial interface was present as an interface to the
implemented control logic. All the RF ports in the system were fully differential.
Output power of 2 dBm was measured in each channel, and the system achieved
more than 20 dB isolation between the antenna and input ports in the receive mode.

Fig. 4.17 Microstrip layout
of a rat-race coupler

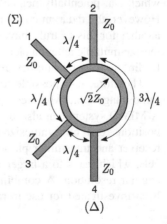

4.8 Integrated Multiple Beam Antennas

To combat multipath and increase the channel capacity of an RF system, two common approaches are to use phased arrays (as has been discussed at length) or multiple beam antennas. The reasonable small aperture of millimeter-wave reflector antennas means that rapid scanning is a possibility, and multiple feeding elements can be used to generate multiple beams. On the other hand, beam scanning can be achieved by simply moving the focal point of the feeding element around. Much of the effort expended toward integrated beamforming networks are related to the substrate-integrated transmission lines discussed earlier in this chapter, because they are so useful at millimeter-wave frequencies [59].

Beamforming networks based on circuit elements use transmission lines to interconnect combiners and couplers. Fixed phasing components provide the phase shifts required for beam scanning [60]. Two main classes of circuit-based beamforming networks exist; these are known as Butler and Blass matrices. The Nolen matrix is a generalized form of the aforementioned matrices.

A Butler matrix is designed to connect 2^n array elements to the same number of beam ports [61]. It is constructed from alternate rows of fixed phase shifters and hybrid junctions. However, crossover sections are inevitable with this configuration, which often result in mismatch, loss, and cross-coupling issues. To overcome these issues, several authors have proposed beamforming networks that are built on substrate-integrated waveguide (SIW) technology [62–65].

4.9 Concluding Remarks

With increasing demands and space constraints in millimeter-wave communication and sensing systems, a definitive shift toward integrated systems is rapidly occurring. To support this process, the SiGe BiCMOS process is becoming commonplace in the development of on-chip systems. As we have seen, newly developed transmission lines— that is, substrate-integrated guides—have played a massive role in monolithic integration and will continue to be a crucial part of integrated system development.

References

1. J. Copeland, W. Robertson, R. Verstraete, Antennafier arrays. IEEE Trans. Antennas Propag. **12**(2), 227–233 (1964)
2. H. Wheeler, Small antennas. IEEE Trans. Antennas Propag. **23**(4), 462–469 (1975)
3. K. Wu, Y.J. Cheng, T. Djerafi, W. Hong, Substrate-integrated millimeter-wave and terahertz antenna technology. Proc. IEEE **100**(7), 2219–2232 (2012)
4. D.M. Pozar, *Microwave engineering*, 4th edn. (Wiley, Hoboken, New Jersey, 2012)

5. K. Wu, D. Deslandes, Y. Cassivi, The substrate integrated circuits—a new concept for high-frequency electronics and optoelectronics, in *6th International Conference on Telecommunications in Modern Satelite, Cable and Broadcasting Service (TELSIKS)*, vol. 1 (2003), pp. 3–9
6. M. Bozzi, A. Georgiadis, K. Wu, Review of substrate-integrated waveguide circuits and antennas. IET Microwaves Antennas Propag. **5**(8), 909 (2011)
7. M. Bozzi, D. Deslandes, A. Paolo, L. Perregrini, K. Wu, G. Conciauro, Efficient analysis and experimental verification of substrate-integrated slab waveguides for wideband microwave applications. Int. J. RF Microw. Comput. Eng. **15**(3), 296–306 (2005)
8. W. Che, C. Li, P. Russer, Y.L. Chow, Propagation and band broadening effect of planar integrated ridged waveguide in multilayer dielectric substrates, in *IEEE MTT-S International Microwave Symposium Digest* (2008), pp. 217–220
9. M. Bozzi, S.A. Winkler, K. Wu, Broadband and compact ridge substrate-integrated waveguides. IET Microwaves Antennas Propag. **4**(11), 1965 (2010)
10. T. Yoneyama, S. Fujita, S. Nishida, Insulated nonradiative dielectric waveguide for millimeter-wave integrated circuits, in *IEEE MTT-S International Microwave Symposium Digest* (1983), pp. 1002–1008
11. K. Wu, L. Han, Hybrid integration technology of planar circuits and NRD-guide for cost-effective microwave and millimeter-wave applications. IEEE Trans. Microw. Theory Tech. **45**(6), 946–954 (1997)
12. Y. Cassivi, K. Wu, Substrate integrated nonradiative dielectric waveguide. IEEE Microwave Wirel. Components Lett. **14**(3), 89–91 (2004)
13. A. Patrovsky, K. Wu, Substrate integrated image guide (SIIG)—a planar dielectric waveguide technology for millimeter-wave applications. IEEE Trans. Microw. Theory Tech. **54**(6), 2872–2879 (2006)
14. A. Patrovsky, K. Wu, Substrate integrated image guide (SIIG)—a low-loss waveguide for millimetre-wave applications, in *35th European Microwave Conference*, vol. 2 (2005), pp. 897–900
15. I. Tekin, M. Kaynak, A 77 GHz on-chip strip dipole antenna integrated with balun circuits for automotive radar, in *IEEE Antennas and Propagation Society International Symposium (APSURSI)* (2012), pp. 1–2
16. Q.L.Q. Lai, G. Almpanis, C. Fumeaux, H. Benedickter, R. Vahldieck, Comparison of the radiation efficiency for the dielectric resonator antenna and the microstrip antenna at Ka band. IEEE Trans. Antennas Propag. **56**(11), 3589–3592 (2008)
17. Q. Lai, C. Fumeaux, W. Hong, R. Vahldieck, 60 Ghz aperture-coupled dielectric resonator antennas fed by a half-mode substrate integrated waveguide. IEEE Trans. Antennas Propag. **58** (6), 1856–1864 (2010)
18. M.H. Nemati, I. Tekin, A 77 GHz on-chip microstrip patch antenna with suppressed surface wave using EBG substrate, in *IEEE Antennas Propagation Society (AP-S) International Symposium* (2013), pp. 1824–1825
19. N. Camilleri, B. Bayraktaroglu, Monolithic millimeter-wave IMPATT oscillator and active antenna. IEEE Trans. Microw. Theory Tech. **36**(12), 1670–1676 (1988)
20. G. Thoren, M. Virostko, A high-power W-Band (90-99 GHz) solid-state transmitter for high duty cycles and wide bandwidth. IEEE Trans. Microw. Theory Tech. **MTT-31**(2), 183–188 (1983)
21. J. Birkeland, T. Itoh, FET-based planar circuits for quasi-optical sources and transceivers. IEEE Trans. Microw. Theory Tech. **37**(9), 1452–1459 (1989)
22. S. Member, Distributed Bragg reflection dielectric waveguide oscillators. IEEE Trans. Microw. Theory Tech. **MTT-27**(12), 1019–1022 (1979)
23. J.W. Mink, Quasi-optical power combining of solid-state millimeter-wave sources. IEEE Trans. Microw. Theory Tech. **34**(2), 273–279 (1986)
24. D.B. Rutledge, Z.B. Popovic, R.M. Weikle, M. Kim, K.A. Potter, R.C. Compton, R.A. York, Quasi-optical power-combining arrays, in *IEEE MTT-S International Microwave Symposium Digest* (1990), pp. 1201–1204

25. R.C. Compton, R.A. York, A 4*4 active array using gunn diodes, in *Antennas and Propagation Society International Symposium:Merging Technologies for the 90's. Digest.*, vol. 3 (1990), pp. 1146–1149

26. A. Stiller, E.M. Biebl, J.F. Luy, K.M. Strohm, J. Buechler, Monolithic integrated millimeter wave transmitter for automotive applications. IEEE Trans. Microw. Theory Tech. **43**(7 pt 2), 1654–1658 (1995)

27. T. Al-Attar, T.H. Lee, Monolithic integrated millimeter-wave IMPATT transmitter in standard CMOS technology. IEEE Trans. Microw. Theory Tech. **53**(11), 3557–3561 (2005)

28. D. Dawn, P. Sen, S. Sarkar, B. Perumana, S. Pinel, J. Laskar, 60-GHz integrated transmitter development in 90-nm CMOS. IEEE Trans. Microw. Theory Tech. **57**(10), 2354–2367 (2009)

29. W. Shin, B.H. Ku, O. Inac, Y.C. Ou, G.M. Rebeiz, A 108-114 GHz 4x4 wafer-scale phased array transmitter with high-efficiency on-chip antennas. IEEE J. Solid-State Circuits **48**(9), 2041–2055 (2013)

30. K.D. Stephan, N. Camilleri, T. Itoh, A quasi-optical polarization-duplexed balanced mixer for millimeter-wave applications. IEEE Trans. Microw. Theory Tech. **31**(2), 164–170 (1983)

31. S.M. Sherman, *Monopulse principles and techniques*, 2nd edn. (Artech House Inc, Dedham, Massachussets, 2011)

32. U. Nickel, Overview of generalized monopulse estimation. IEEE Aerosp. Electron. Syst. Mag. **21**(6), 27–55 (2006)

33. W.P. Du Plessis, Modelling monopulse antenna patterns, in *Saudi International Electronics, Communications and Photonics Conference (SIECPC)*, vol. 2, no. 4 (2013), pp. 1–5

34. S. Raman, N. Scott Barker, G.M. Rebeiz, A W-band dielectric-lens-based integrated monopulse radar receiver, *IEEE Trans. Microw. Theory Tech.* **46**(12), 2308–2316 (1998)

35. E.C.N.E.C. Niehenke, P.S.P. Stenger, T.M.T. McCormick, C.S.C. Schwerdt, A planar 94-GHz transceiver with switchable polarization, in *1993 IEEE MTT-S International Microwave Symposium Digest* (1993) pp. 167–170

36. N. Byer, B. Edward, D. McPherson, S. Weinreb, F. Rucky, J. Sowers, Electronically steered, receive monopulse, active phased array at 94 GHz, in *IEEE MTT-S International Microwave Symposium Digest* (1996), pp. 1581–1584

37. A. Hajimiri, H. Hashemi, A. Natarajan, X. Guan, A. Komijani, Integrated phased array systems in silicon. Proc. IEEE **93**(9), 1637–1654 (2005)

38. C.A. Balanis, *Antenna theory: analysis and design*, 3rd edn. (Wiley, Hoboken, New Jersey, 2005)

39. H. Hashemi, X. Guan, A. Komijani, A. Hajimiri, A 24-GHz SiGe phased-array receiver—LO phase-shifting approach. IEEE Trans. Microw. Theory Tech. **53**(2), 614–625 (2005)

40. A. Babakhani, S. Member, X. Guan, A. Komijani, A 77-GHz phased-array transceiver with on-chip antennas in silicon: receiver and antennas. IEEE J. Solid-State Circuits **41**(12), 2795–2806 (2006)

41. A. Natarajan, A. Komijani, X. Guan, A. Babakhani, A. Hajimiri, A 77-GHz phased-array transceiver with on-chip antennas in silicon: transmitter and local LO-path phase shifting. IEEE J. Solid-State Circuits **41**(12), 2807–2818 (2006)

42. A. Natarajan, A. Komijani, A. Hajimiri, A fully integrated 24-GHz phased-array transmitter in CMOS. IEEE J. Solid-State Circuits **40**(12), 2502–2513 (2005)

43. X. Guan, H. Hashemi, A. Hajimiri, A fully integrated 24-GHz eight-element phased array receiver in silicon. IEEE J. Solid-State Circuits **39**(12), 2311–2320 (2004)

44. A. Valdes-Garcia, S.T. Nicolson, J.W. Lai, A. Natarajan, P.Y. Chen, S.K. Reynolds, J.H.C. Zhan, D.G. Kam, D. Liu, B. Floyd, A fully integrated 16-element phased-array transmitter in SiGe BiCMOS for 60-GHz communications. IEEE J. Solid-State Circuits **45** (12), 2757–2773 (2010)

45. J.W. Rogers, C. Plett, *Radio Frequency Integrated Circuit Design*, 2nd edn. (Artech House Inc, Hoboken, New Jersey, 2010)

46. S. Kishimoto, N. Orihashi, Y. Hamada, M. Ito, K. Maruhashi, A 60-GHz band CMOS phased array transmitter utilizing compact baseband phase shifters, in *IEEE Radio Frequency Integrated Circuits (RFIC) Symposium* (2009), pp. 215–218

47. J.W. May, G.M. Rebeiz, K.-J. Koh, A millimeter-wave (40–45 GHz) 16-element phased-array transmitter in 0.18-μm sige bicmos technology. IEEE J. Solid-State Circuits **44**(5), 1498–1509 (2009)

48. A. Natarajan, B. Floyd, A. Hajimiri, A bidirectional RF-combining 60 GHz phased-array front-end, in *IEEE International Solid-State Circuits Conference: Digest of Technical Papers* (2007), pp. 202–204

49. K. Okada, N. Li, K. Matsushita, K. Bunsen, R. Murakami, A. Musa, T. Sato, H. Asada, N. Takayama, S. Ito, W. Chaivipas, R. Minami, T. Yamaguchi, Y. Takeuchi, H. Yamagishi, M. Noda, A. Matsuzawa, A 60 GHz 16QAM/8PSK/QPSK/BPSK direct-conversion transceiver for IEEE802.15.3c. IEEE J. Solid-State Circuits **46**(12), 2988–3004 (2011)

50. S. Shahramian, Y. Baeyens, N. Kaneda, Y.K. Chen, A 70-100 GHz direct-conversion transmitter and receiver phased array chipset demonstrating 10 Gb/s wireless link. IEEE J. Solid-State Circuits **48**(5), 1113–1125 (2013)

51. E. Fishler, A. Haimovich, R. Blum, D. Chizhik, L. Cimini, R. Valenzuela, MIMO radar: an idea whose time has come, in *Proceedings of the IEEE Radar Conference* (2004), pp. 71–78

52. A.M. Haimovich, R.S. Blum, L.J. Cimini, MIMO radar with widely separated antennas. IEEE Signal Process. Mag. **25**, 116–129 (2008)

53. J. Li, P. Stoica, L. Xu, W. Roberts, On parameter identifiability of MIMO radar. IEEE Signal Process. Lett. **14**(12), 968–971 (2007)

54. F.C. Robey, S. Coutts, D. Weikle, J.C. McHarg, K. Cuomo, MIMO radar theory and experimental results, in *Conference Record of the Thirty-Eighth Asilomar Conference on Signals, Systems and Computers*, vol. 1 (2004), pp. 300–304

55. A. Hassanien, S.A. Vorobyov, S. Member, Phased-MIMO Radar: A Tradeoff between Phased-Array and MIMO Radars. IEEE Trans. Signal Process. **58**(6), 3137–3151 (2010)

56. A. Hassanien, S.A. Vorobyov, Transmit/receive beamforming for MIMO radar with colocated antennas, in *IEEE International Conference on Acoustics, Speech and Signal Processing*, vol. 1 (2009), pp. 2089–2092

57. R. Feger, C. Wagner, S. Schuster, S. Scheiblhofer, H. Jager, A. Stelzer, A 77-GHz FMCW MIMO radar based on an sige single-chip transceiver. IEEE Trans. Microw. Theory Tech. **57**(5), 1020–1035 (2009)

58. C. Wagner, H.P. Forstner, G. Haider, A. Stelzer, H. Jäger, A 79-GHz radar transceiver with switchable TX and LO feedthrough in a Silicon-Germanium technology, in *Proceediings of IEEE Bipolar/BiCMOS Circuits Technology Meeting* (2008), pp. 105–108

59. Y.J. Cheng, P. Chen, W. Hong, T. Djerafp, K. Wu, Substrate-integrated-waveguide beamforming networks and multibeam antenna arrays for low-cost satellite and mobile systems. IEEE Antennas Propag. Mag. **53**(6), 18–30 (2011)

60. P.S. Hall, S.J. Vetterlein, Review of radio frequency beamforming techniques for scanned and multiple beam antennas. IEEE Proc. H Microwaves Antennas Propag. **137**(5), 293 (1990)

61. J. Butler, Beam-forming matrix simplifies design of electronically scanned antennas. Electron. Des. **9**(8), 170–173 (1961)

62. T. Djerafi, K. Wu, A low-cost wideband 77-GHz planar Butler matrix in SIW technology. IEEE Trans. Antennas Propag. **60**(10), 4949–4954 (2012)

63. A.B. Guntupalli, T. Djerafi, K. Wu, Two-dimensional scanning antenna array driven by integrated waveguide phase shifter. IEEE Trans. Antennas Propag. **62**(3), 1117–1124 (2014)

64. Y.J. Cheng, X.Y. Bao, Y.X. Guo, 60-GHz LTCC miniaturized substrate integrated multibeam array antenna with multiple polarizations. IEEE Trans. Antennas Propag. **61**(12), 5958–5967 (2013)

65. Y.J. Cheng, W. Hong, K. Wu, Millimeter-wave multibeam antenna based on eight-port hybrid. IEEE Microw. Wirel. Components Lett. **19**(4), 212–214 (2009)

Chapter 5
Reflector and Lens Antennas

This chapter is subdivided into two sets of antennas that function on similar principles. While the reflector is known for shaping a beam in a particular direction (based on the location of a feed antenna) by using a reflecting aperture, a lens achieves a shaped beam by diffracting the signal from the source antenna.

The use of reflector antennas can be traced back as far as the discovery of electromagnetic propagation in 1888. However, it was not until the explosion of radar development in World War II that the design process behind many familiar reflector shapes was developed.

Following the years after World War II, reflector antennas were in high demand for applications such as long-distance satellite communications, radio astronomy, and increasingly sophisticated high-resolution radar systems. During this time, many of the analytical techniques regarding the optimization of the illumination of a reflector aperture in order to maximize gain, emerged. Reflector antennas are typically capable of providing extremely narrow beamwidths and higher gains than any other single-element configuration. Reflect arrays have also seen substantial development in the last two decades. As the name implies, this antenna is a combination of the most desirable characteristics offered by reflectors and array antennas.

On the other hand, lens antennas have also been thoroughly investigated for millimeter-wave applications. A low profile and excellent beamforming capabilities make these antennas suitable for a number of applications (especially in automotive radar), and the development of substrate-integrated waveguide technology has been an important enabler in extending their operation into the millimeter region.

5.1 Single Feed Reflectors

A parabolic surface (shown in Fig. 5.1) is most often used as a reflector, because it can produce a highly directional pencil beam with very low sidelobe levels. To avoid mechanical steering of the reflector itself (which in many cases can be quite

© Springer International Publishing Switzerland 2016
J. du Preez and S. Sinha, *Millimeter-Wave Antennas: Configurations and Applications*, Signals and Communication Technology,
DOI 10.1007/978-3-319-35068-4_5

Fig. 5.1 Geometry of a
parabolic reflector

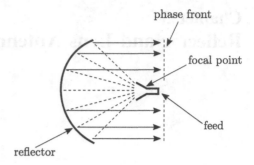

cumbersome), the feed antenna can be moved away from the focal point to produce
a beam that is offset from boresight.

When the feed point is moved away from the focal point, the fields that are
reflected from the aperture are no longer in phase, creating the offset beam.
Figure 5.2 shows some practical examples of reflector antennas.

5.1.1 Feeding Methods

A critical component of the design of a reflector system is selection of the feed
antenna and the decision is largely influenced by the requirements of the system.
For example, a log-periodic dipole feed is advantageous if a very wide bandwidth is

(a)   **(b)**

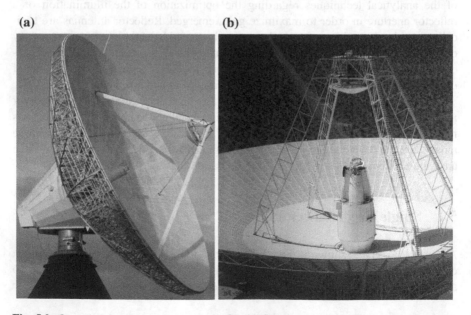

Fig. 5.2 Operational reflector antennas. **a** Dual-reflector satellite communication antenna.
b Space communication antenna used in the NASA Deep Space Network

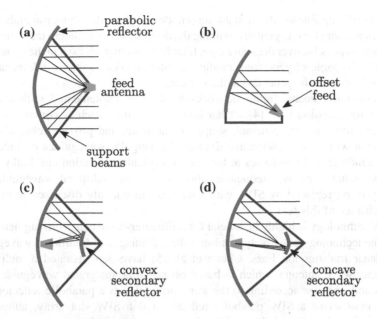

Fig. 5.3 Various methods of feeding a reflector antenna. **a** Axial feed. **b** Offset feed. **c** Cassegrain. **d** Gregorian

desired, and a helix antenna is often used as a feed in some circular polarized systems. However, the horn is by far the most widely used feed because of it provides minimal spillover loss and extremely low insertion loss, together with the ability to produce Gaussian-shaped patterns [1]. Some common feeding methods are shown in Fig. 5.3.

Two types of horn antennas are typically employed, namely the corrugated horn [2] and the dual-mode horn [3]. In the corrugated horn, a regular smooth-walled waveguide is connected to a corrugated waveguide, which can only support the HE_{11} mode. The two sections are matched by gradually varying the slot depth in a short transition region. In the dual-mode horn, two waveguides are also interconnected, one being a dominant-mode circular guide and the second being one of slightly larger diameter. Modes up to TM_{11} are able to propagate via a step transition, and the size of this step is selected such that a precise amount of TM_{11} mode is generated from the TE_{11} mode. The result is that the two modes traveling in the flared horn section that follows the waveguide setup produce equal E- and H-plane patterns.

5.1.2 Beamforming

Beamforming networks have become increasingly complex as a result of the rising demands for multibeam antenna systems. A key problem in the design of beamforming networks is being able to achieve stable amplitude and phase distribution at multiple RF output ports, with the end goal of obtaining the desired radiation pattern that remains as consistent as possible over the scan range. If multiple input ports exist in the system, the network is capable of generating multiple excitation coefficients, which means that multiple beams can be generated in the same array.

Traditionally, beamforming networks have been implemented with different types of transmission lines [4]. Planar transmission lines such as coplanar waveguides and striplines are compact, simple to fabricate, and provide relatively easy integration with active microwave devices. Rectangular waveguides provide high power handling and low losses at the cost of difficult fabrication and bulky structure. At millimeter-wave frequencies, however, most traditional waveguides are steadily being replaced by SIW, which we have thoroughly discussed in the previous chapter of this text.

SIW technology is highly beneficial for millimeter-wave beamforming networks, since the technology effectively combines the advantages of traditional waveguides and planar transmission lines. Cheng et al. [5] have demonstrated a multibeam millimeter-wave antenna, which is based on substrate-integrated waveguide technology and operates according to the same principles as a parabolic reflector. The design consists of a SIW parabolic reflector and a SIW slot array, although a feeding network is included to aid the measurement procedure.

To characterize the performance of the designed beamforming network, the return loss, isolation, and transmitted power were measured and analyzed. At the system center frequency, 37.5 GHz, the return loss was below −15 dB for all ports. When excited at the center port, the total coupling to the remainder of the input ports was measured as less than −20 dB at the same frequency. Several modifications are suggested to lower sidelobe levels. Adding dummy ports, which are terminated with matched loads, is one option, although this does degrade radiation efficiency. Another option is to split each port in the feed side into two, or alternatively combining the split ports with a magic-T, where the difference port is terminated.

5.1.3   Focal Array Fed Reflectors

Reflector antennas fed by focal arrays have been considered for numerous designs and applications that require multiple spot beams, e.g., satellite communications. Such an antenna typically comprises either a single or double reflector surface, illuminated by a dense focal array. A compact horn antenna is a suitable choice of element for the focal array. Nguyen et al. [6] were among the first groups to study the implementation of the transmission of focal arrays through purely dielectric focusing systems. The group focused on millimeter-wave automotive radar applications for their design, although the demonstrated system was designed for 28 GHz operation in order to ease the prototyping process. In these systems, both long- and short-range measurements are used to provide useful data to the driver and increase road safety. As a result, reconfigurable apertures have become an attractive solution to comply with the requirements of such a system. Increased angular resolution, tunable beamwidth, and multiple directive beams are some of the reconfigurable parameters.

The linearity of Maxwell's equations dictate that whenever several elements in an array are driven with the appropriate phase and amplitude coefficients, the beam that results from the focusing system is a vector sum of the individual off-axis beams. Therefore, it is possible to alter the radiation characteristics of the lens by altering the phase and amplitude coefficients of the feeding elements. This is an alternative to optimizing the shape of the lens in beamforming applications, meaning that the lens used in this technique is inherently very flexible.

For analysis and design, the antenna is modeled as a hemispherical lens, illuminated by a focal array. The lens is coated by a matching layer, a quarter of a wavelength thick. By adding this matching layer, the effects that uncontrollable internal reflections have on the desired radiation pattern are minimized. The lens is constructed from Rexolite ($\varepsilon_r = 2.53, \tan \delta = 5 \times 10^{-3}$), and its diameter measures 85.6 mm, while its extension length is 85.6 mm. The values of these parameters were chosen to produce a directivity of approximately 28 dBi. The elements of the focal array were modeled as probe-fed microstrip patch antennas, and it is expected that an aperture-coupled feed would produce similar results. The signal source feeding each element is separately matched and the chosen optimization parameters were the generator amplitude and element spacing.

Several experimental arrays were constructed to validate the theoretical predictions. Square patches that are fed through aperture coupling were used as the array elements in all cases, and these were etched onto 0.254 mm RT/Duroid 5880 substrates. Three different primary sources were experimented with: a single microstrip antenna, to linear H-plane arrays with six elements, and then with ten elements. The prototypes were slightly modified from the numerical model, in the sense that the central patch was replaced by two coupled patches and the matching layer on the lens was removed. The measured cross-polarization was at least as low as −23 dB across the specified bandwidth of 26–30 GHz for all three prototypes. One observation was that the total loss in the beamforming network was larger when more elements were used in the feed array. This was due to the increased losses caused by internal reflections when the feed point was offset from the center of the lens. The ten-element array thus exhibited the largest total loss of 5.7 dB, compared to 2.7 dB for the single-element feed.

The achievable antenna efficiency is greatly affected by the loss in the beamforming network. A directivity of 14.2 dB with a 9.2 dB gain was observed for the ten-element array. When compared with the 28 dB directivity and 25.3 dB gain observed with the single-element feed, it is clear that the efficiency is severely deteriorated. The authors predicted that using high-quality substrate materials (such as quartz) and improved feeding lines (such as stripline or coplanar waveguides) would reduce the total line length, and thus the total loss, by a factor of at least two.

5.2   Reflectarrays

As mentioned in the abstract section, a reflectarray is an antenna built on a combination of the most desirable properties of both arrays and reflector antennas. A simple microstrip reflectarray consists of a planar array of patch or perhaps dipole elements, printed on an electrically thin dielectric substrate [7]. Similar to the familiar reflector antenna, a feed antenna is then used to illuminate the array, where the array is designed such that its individual elements scatter the received wave with a phase angle that will result in the desired phase front. This concept is almost identical to the classical reflector antenna, where the phase front can be offset in angle by moving the feed antenna away from the focal point. As a result, a reflectarray can be regarded as a flat reflector.

While the reflectarray as a concept has been around for some time, it is the rapid development of microstrip antennas, together with the ever-increasing requirement for low-profile, high-gain antennas that led to many reflectarray designs using microstrip elements. As discussed in several sections of this text, losses in the feed network are the major limiting factor when scaling microstrip arrays up to millimeter wavelengths, more so as the arrays become larger. Therefore, when the goal is to obtain high-gain millimeter-wave antennas where the prominent profile of a parabolic reflector is undesirable, a reflectarray is a suitable solution.

A great deal of flexibility is offered by microstrip reflectarrays. Feeding the array can be implemented through offset feeding, cassegrain feeds or prime-focus feeding. It is also possible to scan the main beam and generate monopulse antenna patterns. Circular and linear polarization is obtainable and dual-polarization configurations are also a possibility.

Realizing the phase scattering from individual elements is a major part of the design process, which can be achieved through various methods. One solution is to use an offset feed approach, where the feeding stubs at each microstrip element are made progressively longer [8–10]. Using elements of variable sizes is an alternative (and improved) approach, as demonstrated by Pozar [11, 12]. Both these methods may be considered as introducing a shift in the resonant frequency of each individual element, which then causes the phase of the reflected field to change. However, by using variably sized elements, the designer is granted greater freedom in the layout of the array to achieve different polarizations, together with larger attainable bandwidth. This is contrary to the limitations imposed by stub-tuned elements. Furthermore, similar to conventional microstrip element design, polarization selectivity can be obtained by using circular patches or crossed dipoles to generate circularly polarized waves, or rectangular microstrip patches to generate linear polarization [13].

In terms of bandwidth, a microstrip reflectarray strikes a balance between microstrip arrays (which are limited to low percentage bandwidths) and parabolic reflector antennas (which can achieve very wide bandwidths). This is typically the case, as most of the electrical characteristics of a reflectarray are bound to be inferior to those of a traditional reflector. However, as a result of the polarization selectivity that can be implemented with a reflectarray, the cross-polarization radiation is generally better than what could be obtainable with rectangular patches. Pozar presented an in-depth study on microstrip reflectarrays for millimeter-wave operation. Issues such as tapering efficiency, feed network topologies, material loss and pattern bandwidth are discussed in great detail [7]. Four experimental reflectarray designs were constructed for operation at 28 and 77 GHz. The last two variations are of particular interest here. One of the designs used a prime-focus feed and with the other a cassegrain feed. Experimental measurements revealed a beamwidth of 1.55° × 1.65° (which corresponds to a directivity of approximately 41 dB) and a peak sidelobe level of 25 dB, where a prime-focus flared waveguide J-feed was used.

The cassegrain-fed reflectarray had a beamwidth of 1.5° × 1.7°, with a measured gain of 39.6 dB and sidelobe levels peaking at 19 dB.

5.2.1 Dual Reflector

A well-known dual reflector is the classical Cassegrain antenna, which uses a hyperbolic aperture for the subreflector and a parabolic contour for the main reflector [14, 15]. The Cassegrain antenna illustrated earlier is shown again in Fig. 5.4.

This creates two focal points in the system. One of these, the real focal point, is at the phase center of the feed element and the other, the virtual focal point, is located at the focal point of the main reflector. The efficiency of a dual reflector is dependent on the ability to illuminate only the reflector apertures and the uniformity by which the parabola is illuminated. These parameters are, respectively, known as spillover efficiency and illumination efficiency [1].

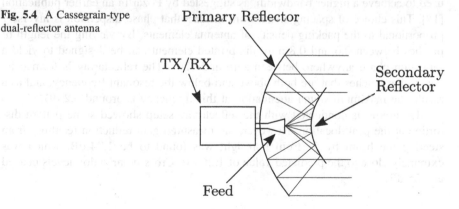

Fig. 5.4 A Cassegrain-type dual-reflector antenna

A 94 GHz dual-reflector antenna, which uses a reflectarray as subreflector, has recently been demonstrated by Hu et al. [16]. Dual reflectors are often encountered in remote sensing applications such as earth observation. The main reflector is a familiar parabolic reflector, which is fed from a printed reflectarray, which in turn is fed from a W-band pyramidal horn. The antenna produces a main beam that is 2.4° wide at the 3 dB point and the measured gain peaks at 38.7 dB at 94 GHz. The parabolic reflector is a large machined chunk of aluminum alloy and the surface was polished to produce a 4 μm surface roughness. The diameter of the main reflector measures 120 mm. Two configurations were studied by Hu et al. First, three flat metal subreflectors were positioned at set angles in order to produce a main beam at azimuth angles of 2.5° and 5°. Thereafter, a passive reflectarray was used to tilt the beam away from broadside by about 5°. This was achieved by varying the sizes of the microstrip patches that formed the reflectarray, and the result was a predetermined phase distribution across the aperture of the subreflector. The microstrip array consisted of 28×28 elements printed on a 115 μm conductor-backed quartz wafer with $\varepsilon_r = 3.78$, $\tan \delta = 0.002$.

The far field patterns were measured with a dual-channel receiver with a millimeter-wave down-conversion module. This converter drives W-band harmonic mixers that are placed in both the reference and test channels, together with an 11.45 GHz phase locked oscillator signal. The result is that the 94 GHz signal is mixed down to a 2.4 GHz IF, which is then passed through a preamplifier before it is mixed down a second time. The heterodyne configuration described here provided the setup with 60 dB of dynamic range when the standard gain horn was used as a source, and the source was configured to provide 20 dBm output power. The measurements were done in an anechoic chamber with a distance of approximately 7 m between the two points, corresponding to $1.6 \, D^2/\lambda$. Other than a sidelobe level of 23 dB, which was slightly higher than determined in the simulated system, the theoretical and practical results were in good accordance with one another.

To validate the proposed beam-scanning technique, the reflector plates were replaced by the designed microstrip reflectarray. An element spacing of 0.3λ was used to achieve a higher bandwidth, as suggested by Pozar in an earlier publication [17]. This choice of spacing stems from the fact that phase dispersion is inversely proportional to the packing density of antenna elements. By varying the length of patches between 0.6 and 0.8 mm, the printed element can be designed to yield a reflection phase anywhere between zero and 320°. The reflectarray is formed by combining patches that are both above and below the resonant frequency, and as a result, the maximum signal attenuation at this frequency is around 0.2 dB.

Measurements conducted with the reflectarray setup showed some pattern distortion at the peak location. However, the measured gain reduction resulting from steering the beam by 5° from boresight was found to be 1.24 dB, which was extremely close to the predicted value of 1.26 dB. Cross-polarization levels peaked at −25 dB.

5.2.2 Reconfigurable Reflectarrays

In applications such as terrestrial communications, remote sensing, radar and microwave imaging, electronically reconfigurable reflectarray antennas may be beneficial and such antennas have been studied intensely in recent years. In order to implement the reconfigurable functionality, solid-state tuning devices such as FETs or PIN diodes are typically used. Recent developments in RF micro-electromechanical switches (MEMS) means that these devices may offer different possibilities as a new class of tuneable devices, as opposed to semiconductor based devices, which are inherently limited by losses and nonlinearity. Several groups have intensely studied the use of MEMS devices in reconfigurable reflectarrays. Hum, McFeetors and Okoniewski reported on a tuneable reflectarray using MEMS capacitors [18]. The concept is comparable to a tuneable parallel plate capacitor and this has been realized through either a bridge or a cantilever design. However, the bridge design provides improved tuning stability in comparison. The resulting antenna system showed greatly reduced intermodulation distortion and losses when equated to semiconductor tuning devices used in the past.

As we have mentioned, reflectarrays are a particularly attractive option at millimeter-wave frequencies, mainly because traditional microstrip radiators exhibit intolerable feed network losses at these frequencies.

In a review paper by Sorrentino et al. [19], important developments in millimeter-wave reconfigurable reflectarrays since the turn of the century are investigated. Thus far we have discussed, in detail, how phase shifts are synthesized by varying the shape of reflecting elements. To obtain a reconfigurable design in this case, the tuning device can be integrated with the radiating element. Menzel has demonstrated designs that are reasonably compact, using polarizing grids as the subreflector element [20, 21]. Liquid crystals have been used to implement reconfigurable functionality, since their dielectric characteristics are voltage-dependent [22, 23]. Another technique is centered on using varactor diodes, as demonstrated by Hum et al. [24].

5.2.3 Folded Reflectors

Printed and folded reflector antennas that are quasi-optically fed are an alternative solution for printed, low-cost antennas. The configuration of such an antenna consists of an array of printed antennas (which act as phase shifters with a fixed reflection) and a printed polarization diplexer, such as a slot array or a printed grid. The baseline of this antenna is the microstrip patch array, printed on a grounded dielectric substrate. Receiving a plane wave from broadside will result in the complete power being reflected, but the phase angle of the reflected wave depends on the patch dimensions. These patches operate far from resonance, which is in contrast to traditional microstrip arrays where resonant length patches are used

Fig. 5.5 A folded reflector antenna

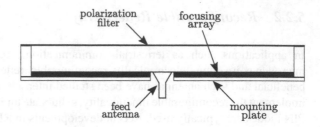

solely as radiating elements and phase shifting is done one way or another in the feed network.

Consequently, without the additional transmission lines involved in the feed network, these antennas exhibit much lower losses. Therefore, independently choosing the dimensions of the patch elements creates the possibility of realizing different properties for the two polarizations, such as a dual-band antenna. The focusing array can be altered to *twist* the polarization of the electromagnetic field, which in combination with a polarizing grid, creates a folded reflector antenna [20].

A folded reflector consists of a feed, a printed reflectarray and a polarization filter, as shown in Fig. 5.5. In terms of the feed antenna, a cylindrical-waveguide horn is commonly employed, but a planar array can be equally effective. The polarization filter may be either a grid or a resonant slot array and it also acts as a radome in this configuration. The function of the polarization filter is implied by the name; it is transparent for waves of one polarization and reflects waves of another polarization.

The polarization of the feed antenna is chosen so that the generated wave is reflected by the printed slot array (or grid) toward the focusing array. The design of this reflectarray has been intensely studied by Menzel et al. [20], among others.

Relative to the incident electric field, the patch axis is rotated by 45°. Since the electric field can be decomposed into two components that are parallel to the patch axis, their reflection characteristics may be analyzed individually. The patch dimensions are specified to lead to a 180° difference in phase between the two field components. The printed reflector was designed with a circular aperture. Both the reflectarray and the polarizing grid were fabricated on an RT/Duroid substrate, with dielectric constants of 2.2 and 2.5, and thickness of 0.254 and 1.58 mm, respectively. The feed antenna was chosen as a circular-waveguide horn with a diameter of 5.5 mm.

Narrow beamwidths of 3.2° and 3.4° in the E- and H-planes, respectively, were found, and the highest sidelobe level was −24 dB at a design frequency of 61 GHz. Over the measurement bandwidth, the sidelobe level was found to peak at −20 dB in both the E- and H-planes. The measured gain was 34 dB at its maximum, and a 3 dB gain reduction occurred at a 7 GHz bandwidth.

A second set of measurements was conducted on a variation of this antenna, where the reflector and polarizer were moved closer to each other, such that the total depth of the antenna did not exceed 15 mm. A slight degradation of performance was observed, with the peak gain reducing to 31.5 dB and sidelobe level reaching a maximum of −22 dB in the H-plane.

Furthermore, the authors also designed a 76.5 GHz antenna with a movable reflector, in order to facilitate beam steering. The aperture diameter was 100 mm and a 25 mm spacing between the reflector and polarizer was chosen. The slot array was printed on a 1.02 mm substrate with $\varepsilon_r = 4.5$, to enable improved performance in terms of polarization, although slightly higher losses were expected with this substrate. Up to a scan angle of 6°, the measured sidelobe level remained below -20 dB and the antenna gain was observed as 35 dB at the design frequency.

5.2.4 Retrodirective Arrays

A retrodirective array can be regarded as a special subset of reflector (and reflectarray) antennas [25]. It is designed to reflect incident signals back toward the source, without knowledge of the location of the source beforehand. This is quite a unique characteristic and it makes retrodirective antennas suitable for applications such as radar [26–28], electronic warfare [29–31], and communication systems [32, 33]. Perhaps, the most well-known retrodirective array is known as the van Atta array [34], where elements of a symmetric array are connected by equal-length transmission lines, as shown in Fig. 5.6. These arrays are challenging to implement in practice, because of strict symmetry and phase-matching requirements. To overcome this, phase conjugation has been implemented with heterodyne mixing [27, 35].

Retrodirective arrays have been used for millimeter-wave systems on a number of occasions. The first of these was in 2002, when Buchanan, Brabetz and Fusco reported a 62/66 GHz frequency offset retrodirective array [36]. The separation of center frequencies between transmit and receive modes is done to improve isolation between the two signals. The system consists of several active components fabricated on a Taconic TLY substrate using the OHMMIC ED02AH process. The operation of the antenna can be described as follows: The receiving antenna captures a 65.5 GHz signal, which is then mixed down to 62.5 GHz using a set of 3 GHz LO

Fig. 5.6 Diagram of a passive van Atta retrodirective array

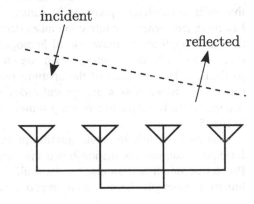

signals. The signal is then amplified and retransmitted at 62.5 GHz. The transmit antennas are crossed over, so that their phase relationships allow the retransmitted signal to be steered by the source. This approach also provides the freedom to scale the element spacing to the center frequency, reducing the beam-pointing error that results from the offset frequency.

Ren and Chang have also presented their retrodirective array, which is designed for K_a-band applications and provides a 33.7 % 10 dB bandwidth between 31 and 42.8 GHz [37]. This design involves an 8×16 array of proximity coupled dual-ring antennas, and 4×4 subarrays are built in a retrodirective configuration. This is a two-layer antenna printed on RT/Duroid 5880 substrates with thickness 0.79 and 0.51 mm, and two TM_{110} modes are used in an attempt to widen the bandwidth.

Ali et al. [38] used the SIW to build a 30 GHz passive van Atta array. The array design is based on alternating SIW slot arrays, where the longitudinal slots are etched in a ground plane covering the entire circuit. An interesting design has recently been demonstrated by Christie et al. which is in the form of a retrodirective array based on the Rotman lens [39]. The beamforming network exhibits a conjugate phase response when the input ports are terminated in open or short circuits, which are exploited to achieve an automatic target-tracking function. In order to compare the experienced RCS enhancement, a 12-element Vivaldi array was constructed with and without the Rotman lens, and measured results revealed a significant increase in RCS when the lens was included in the configuration.

5.3 Lens Antennas

Lens antennas are a primary candidate for the reduction of surface wave losses. Being a subset of dielectric antennas (such as dielectric resonators, rods and horns), lenses can be subdivided into several categories based on their constitutive materials and shape. Fresnel zone plate lenses are a common example of such an antenna built from an inhomogeneous material, since the lens is built with dielectric rings that each has different permittivity values. Another inhomogeneous lens is the Luneburg lens, where the refraction index decreases from the center [40]. In spite of this, most applications make use of homogeneous lenses, often combined with a type of anti-reflection coating when the material dielectric constant exceeds a particular value. In terms of the aperture profile, lenses can either be shaped or canonical. Shaped lenses are typically designed to produce a desired radiation pattern, which is unique to a primary source. Typical lens configurations are shown in Fig. 5.7.

It is also possible to create multibeam antennas with acceptable off-axis performance. Rotman has demonstrated this with an aplanatic-zoned lens [41], and Peebles accomplished similar results with a dielectric bifocal lens [42]. At millimeter wavelengths (and in some reports, submillimeter wavelengths), canonical

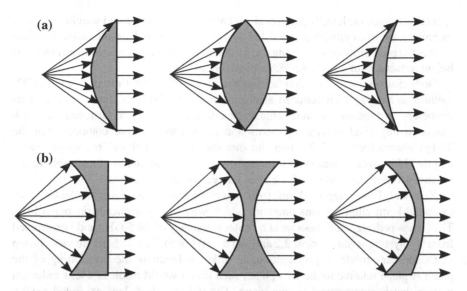

Fig. 5.7 Lens configurations with **a** refraction indices n > 1 and **b** refraction indices n < 1 [13]

shapes have been extensively studied; see for example reports on the circularly polarized substrate lens by Wu et al. [43], as well as the modified rectangular loop slot design by Otero et al. [44]. The low profile of lens antennas makes them highly desirable for automotive radars, where space constraints play a crucial role in the design of the antenna [45].

5.3.1 Reduced Size Lens

The first detailed investigation on reduced size lenses (those with diameters less than a few wavelengths) was conducted by Godi et al. [46]. The aim of research in this field is to establish a design theory for compact radiators with directivities that range from 10 to 25 dB, which exhibit low surface wave losses and high-radiation efficiencies. To feed the lenses under investigation, aperture-coupled patch elements have been used. A challenging aspect of this work is that previously established theory was developed from analyzing electrically large lenses, which are several free space wavelengths in diameter. More sophisticated analysis techniques have since been developed for electrically small lenses.

Internal reflections affect the radiation characteristics of lenses, and these effects have been thoroughly investigated and documented. Accounting for first order reflections is sufficient when computing the radiation pattern for materials with a permittivity of below 4. Higher order reflections spread their power out in all directions, without having a detrimental effect on the desired main lobe. Furthermore, the input impedance and mutual coupling are affected by the occurrence of multiple reflections inside the lens. Nguyen et al. recently proposed a size and weight reduction technique, using a cylindrical, air-filled cavity. An additional air–dielectric interface creates the possibility of shortening the focal length, and in this case led to weight and height reductions of 27 and 13 %, respectively [47].

Godi, Sauleau, and Thouroude used the finite-difference time domain (FDTD) method to analyze their reduced size lens in the 47–50 GHz band. Various lens diameters and materials were characterized, and the quartz structure $\varepsilon_r = 3.8$ exhibited improved radiation patterns and proved to be more compact than the Teflon alternative $\varepsilon_r = 2.2$ when the diameter exceeded three free space wavelengths. Moreover, it was shown that when the lens diameter is comparable to the system wavelength, surface efficiencies can easily exceed 200 %.

Xue and Fusco reported on their development of a planar dielectric slab extended hemi-elliptical lens antenna, which was fed by a microstrip patch [48]. The lens was designed to operate at a center frequency of 28.5 GHz and constructed from a polystyrene sheet ($\varepsilon_r = 2.2$ and $\tan \delta = 0.0005$). The TM_0 mode was chosen as the primary mode of propagation, in order to facilitate the positioning of the patch H-plane relative to the lens plane, such that it would result in a lens radiation pattern that is symmetrical in this plane. The 6.4 mm thick lens exhibited axially symmetric patterns with an 18.5 dB gain at 28.5 GHz, with 40° E-plane and 4.1° H-plane 3 dB beamwidths, as well as a 10 % impedance bandwidth (for $S_{11} \leq -10 \, \text{dB}$). In the multibeam mode of operation, the lens was demonstrated to launch up to nine beams with a total scan range of approximately 27°, while suffering a maximum of 2.8 dB loss in gain for off-axis beam positions. The low profile of the antenna makes it suitable for a number of applications, such as broadband wireless systems and vehicular telemetry.

Another lens, developed for 60 GHz mobile broadband applications, has been reported by Rolland et al. [49]. The antenna consists of a flat dielectric lens that is designed to produce an omnidirectional E-plane pattern, and a flat-top beam in the H-plane. The geometrical optics/physical optics method that has often been used to study lens shapes was not exactly applicable to the flat-shaped lens developed. Therefore, the authors developed a two-stage analysis procedure in order to optimize the flat lens. For the first stage of the optimization, a 2D CAD tool was used, which combines a real-valued genetic algorithm and 2D FDTD solver to obtain an optimized shape, based on a predetermined radiation pattern. This approach is advantageous for several reasons. It enables rapid analysis of the antenna over a wide bandwidth with a single execution cycle. It reduces the complexity of the implementation, provides highly accurate numerical computations and is compatible with any material and shape.

The second stage consisted of transforming the 2D solution into a 3D solution, starting with replacing the 2D numerical feed previously used to a 3D feed, which exhibits radiation properties that are similar to those of the 2D feed. Optimizing the thickness of the flat lens is also a two-part procedure, depending on the particular configuration in question. In the first case, the two flat-sides of the lens structure are covered with perfect electric conductor (PEC) walls. For an E-plane structure, where the electric field is parallel to the two PEC planes, a permittivity correction is required in order to ensure that the value used is in accordance with the 2D model. This is not needed for H-plane configurations, where the electric field is orthogonal to the PEC planes. In the second case, the flat sides of the lens are directly in contact with air, resulting in a lower effective permittivity, relative to that of the lens material. Lotspeich has detailed the required permittivity correction values for a number of configurations [50].

An experimental setup was constructed and measured in the 57–63 GHz band. H-plane radiation patterns exhibited a sidelobe level of less than −12 dB, with a maximum pattern ripple of 5 dB, which is indicative of a good quality flat-top beam. Overall, the predicted antenna performance was in excellent accordance with the measured results. The input reflection coefficient was measured at below −14 dB over the entire bandwidth, with an average directivity of about 10 dBi.

5.3.2 Rotman Lens

A Rotman lens is often used in millimeter-wave beamforming applications. While it does not offer the wide bandwidth or large scan angles of other types of beam-forming lenses, it provides a compromise in performance and performs well in terms of scan range, sidelobe levels and efficiency. The structure comprises a parallel plate region, with waveguide ports placed around the edges of the plate regions, as shown in Fig. 5.8. These ports are then fed by an array of switches, each connected to a radiating element. Power input to a particular focal port will produce a beam in a given direction and steering is achieved electronically by switching the

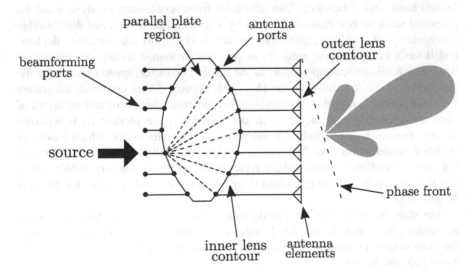

Fig. 5.8 Principle of operation for a Rotman lens

input from one focal port to the next. This enables multiple simultaneous beams and an application example would be an amplitude comparison monopulse system.

Although Rotman patented his idea in 1965 [51, 52], utilizing the properties of the Rotman lens for millimeter-wave applications was not explored until the 1990s. Two papers on the subject were presented at the 1997 IEEE Radar Conference, as a collaboration between the Georgia Institute of Technology and the U.S. Army Research Laboratory.

Both designs focused on a Rotman lens beamforming network designed for K_a-band operation (33–37 GHz) [53, 54]. The design consisted of 32 antenna ports, 17 beam ports, and a maximum achievable scan angle of $\pm 22.2°$. Slight alterations to the lens design can enable scan angles of up to $\pm 60°$, at frequencies in the 94 GHz range. Measurements were performed on a prototype system, which revealed a maximum sidelobe level of -30 dB and an insertion loss of at most 2.3 dB.

Two lens designs in the upper millimeter-wave region were recently demonstrated. Nüßler presented his 32-element Rotman lens antenna, designed for operation at 220 GHz, at the 2015 German Microwave Conference [55]. Nüßler highlighted the primary difference between low frequency and millimeter-wave Rotman lens designs, in that the parallel plate region and the waveguide lens are integrated into a single-plane block. Millimeter-wave modules from Oleson were used to extend the measurement range from 140 to 220 GHz, and a 67 GHz Keysight general-purpose network analyzer (PNA) was used to characterize antenna performance. Sidelobe performance of at least -15 dB was achieved over a $\pm 40°$ scan range and the amplitude loss for off-axis scanning was negligible.

Jastram and Filipovic used micromachining technology to create their W-band lens design [56]. Achieving ultra-wide bandwidths that are possible with Rotman lenses is a challenging task, mainly because of the limitations of transmission lines. Rectangular waveguides have low losses, but this is offset by their bulky nature and limited bandwidth. Microstrip lines also suffer from very limited bandwidth and the proposed solution is a Rotman lens fed by a surface micromachined double-ridge waveguide. The PolyStrata fabrication process was used to micromachine the lens, and it has a unique nature, since it can produce air-loaded rectangular microcoax lines with a dielectric strap support as the center conductor, greatly increasing the bandwidth of the transmission line [57]. This approach also simplifies integration with components and other waveguides. The process consists of adding layers of photoresist, copper, and dielectric straps. Thereafter, the photoresist is removed through release holes periodically inserted into the outer walls, which leaves an air-filled transmission line. The number and height of these layers can be varied between fabrication sessions and are typically preset by the foundry, which means that the lens can be further optimized if a wafer with the design values for the layer heights is used.

The dimensions of the waveguide ridges (as shown in Fig. 5.9) are chosen to widen the single-mode bandwidth. For Filipovic and Jastram's design, the dimensions in millimeters were as follows: $a_1 = 0.8$, $a_2 = 1.65$, $b_1 = 0.8$, $b_2 = 0.35$ and $b_3 = 0.2$.

Fig. 5.9 Double-ridged
waveguide design parameters

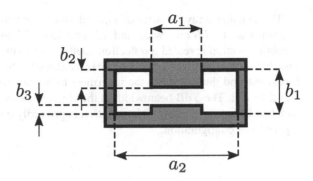

The transmission line maintains a maximum attenuation of about 0.11 dB/cm over its length and provides single-mode operation from 55 to 175 GHz. A key aspect of the lens design is the transition used between the double-ridged waveguide and the parallel plate in the Rotman lens. Knott proposed a solution to this problem by introducing an exponential taper in the top and bottom walls of the waveguide and widening it laterally [58]. However, the discrete values to which the foundry process is limited make it unfeasible to insert a smooth exponential taper and a step approximation was rather implemented. Five steps were used for the top wall and three for the bottom wall and the waveguide width was extended to 2.6 mm (from the original width of 1.65 mm). To improve the performance of the waveguide further, the final step in the ridge was extended into the parallel plate cavity by 0.5 mm. This type of transition provides insertion loss below 0.75 dB and a reflection coefficient of below −10 dB, over a bandwidth of 70–170 GHz.

The geometrical optics (GO) approach is typically used in lens designs. It assumes the absence of multipath interference, a lossless lens, and that all the energy fed into a beam port is coupled to the array contour. These idealizations are of course not present in practice, and they must be accounted for in the design process. For a topology using three focal points, three points exist on the beam contour where one can expect no path length errors. At all other points along the contour, deviations occur because the path length is no longer a function of the beam position. The lens presented by Filipovic and Jastram is a 4 × 5 Rotman lens with a focal length of 12, 2 mm spacing between array ports and a maximum focal angle of ±30°. The system center frequency was chosen as 100 GHz.

In order to guarantee that the photoresist layer is removed from the parallel plate region, 250 μm wide square release holes are inserted at 500 μm periods. The minimum size of these holes is determined by the chemical process used. While larger holes ease the photoresist removal process, they can degrade the electrical performance. To investigate the impact of hole sizes, the lens was simulated over a 60–170 GHz bandwidth. At least 32 % of the input power is coupled to the array ports, corresponding to a 5 dB insertion loss, and there was no observable radiation losses with the addition of release holes. The array ports are then connected to the array elements through the coaxial lines described earlier, phased as per recommendations by Rotman and Turner [51].

The antenna array consists of tapered slots to ensure wideband operation and measures up to 7.6 mm wide and 7.7 mm long. Measurements on the designed transition section revealed a reflection coefficient below −7.5 dB, with an insertion loss below 1.5 dB over a 65–140 GHz bandwidth. Several beam positions were measured, and the gain from the innermost to the outermost beams varied between 5 and 13 dBi. The 3 dB beamwidth in the E-plane varied between 40° and 20° at 60 and 140 GHz, respectively This variation is primarily attributed to the nature of the tapered slot configuration.

5.3.3 Lüneburg Lens

Lüneburg lenses (illustrated in Fig. 5.10) have been developed to create large-diameter, highly efficient lens antennas. It is a gradient-index lens that is spherically symmetric and similar to the Rotman lens, it is often sandwiched between two parallel plates to create the antenna.

Hua et al. demonstrated a 30 GHz modified Lüneburg lens antenna in 2012 [59]. The spacing between the parallel plates was varied along with the normalized radius, in order to obtain a general Lüneburg's variation of the effective refraction index. The chosen feed antenna was a planar antipodal linearly tapered slot antenna, and it was placed between the parallel plates at the focal point of the lens. The antenna was fabricated on an Arlon Di 880 substrate, 0.254 mm thick, with $\varepsilon_r = 2.2$. The antenna generated a fan beam and the measured 3 dB beamwidths were 8.6° and 68° for the E- and H-planes, respectively. E-plane sidelobe levels peaked at −20 dB and the measured cross polarization was at most −28 dB. Furthermore, over a 4 GHz bandwidth between 28 and 32 GHz, the antenna efficiency varied between 50 and 71 %, while the measured gain at the design frequency was 16.6 dBi, corresponding to an efficiency of 68 %.

This group further investigated this antenna by developing multiple beam capabilities [60]. This was accomplished by implementing the slot array antennas in an arc at the periphery of the lens, and switching between the excited elements in the array. Eleven beam positions were measured and it was concluded that the scan range reached ±80° at best. This was the result of large attenuation observed when

Fig. 5.10 Basic structure of a Lüneburg lens antenna

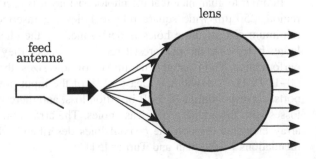

exciting elements at the outer edges of the array, which was to be expected, seeing that blocking effects become more apparent with an increase in element number. However, the maximum sidelobe level did remain below −15 dB over a scan angle of ±86°.

Lafond et al. developed a 60 GHz antenna based on the Lüneburg lens, which offered a reconfigurable radiation pattern [61]. The permittivity distribution in the lens aperture was controlled by means of two different hole densities near the center and also by altering the dielectric thickness of the edge. This inevitably required drilling into the Teflon material, which could lead to some inaccurate results at 60 GHz. The design presented by Lafond et al. used an open-ended ridged waveguide, which was fed by a microstrip line. A step change in waveguide height was used to match the impedance at the waveguide-to-microstrip transition. The microstrip line was printed on a 0.127 mm Rogers RT/Duroid 5880 substrate, with $\varepsilon_r = 2.24$ The simulated reflection coefficient was below −9 dB over the specified bandwidth (59–64 GHz).

The source antennas are relatively narrow at 2 mm each, and it is thus possible to place many of these on the plate edges. If 16 of these sources are used, each providing a scan angle of about 10°, a total scan area of ±75° can be achieved. In order to reconfigure the radiation pattern, a set of monolithic microwave integrated circuit (MMIC) amplifiers and a circular power divider are used to excite all the waveguide ports in phase. Beam steering is done by switching a particular source on, therefore creating a beam that is offset from the normal position. Furthermore, by switching several sources simultaneously, it is possible to alter the radiation pattern. This can be difficult to accomplish for wideband systems, especially considering that the antenna must be suitably matched, irrespective of the number of sources.

To overcome this, a wideband power amplifier was used, which serves a double purpose as a switch, providing a matched connection in both the 'on' and 'off' states. This greatly simplifies the implementation of the feeding network. One of these amplifiers was placed behind each waveguide to provide control over the excitation. With the amplifiers in place, the lens antenna was manufactured by connecting the MMICs to single-layer capacitors using wire bonds. Measurements on the experimental system correlated well to the simulated system, and several beam positions between −55° and 35° were demonstrated, although some issues with the amplifiers invalidated beam measurements beyond 35°.

5.4   Concluding Remarks

Reflector antennas, being one of the earliest developed structures, have undergone significant variations, improvements, and alternate configurations. In order to adapt to the increasing demands of applications that demand low-profile antennas, traditional millimeter-wave reflectors have been phased out and replaced with new compact structures that still provide acceptable performance. Reflectarrays are perhaps the first example of this, aiming to achieve a balance between the low-profile nature of microstrip arrays and the extremely narrow beamwidth and high gain achievable by conventional parabolic reflectors.

Beam steering with reflector antennas can be cumbersome, since either the reflecting surface or the feed antenna needs to be mechanically moved. In this regard, lens antennas are a very useful component, even more so because of their capability to scale so well into even upper millimeter-wave bands. Lenses also lend themselves well to integrated designs and multiple beam antennas, since beam steering is accomplished largely through the clever design of the feed network. These attributes make a variety of lens configurations very attractive options for a number of millimeter-wave applications.

References

1. D.R. Jackson, A.A. Oliner, *Modern Antenna Handbook* (Wiley, New York City, New York, 2008)
2. B. Maca, Design of corrugated conical horns. IEEE Trans. Antennas Propag. **AP-26**(2), 367–372 (1978)
3. H.M. Pickett, J.C. Hardy, J. Farhoomand, Characterization of a dual-mode horn for submillimeter wavelengths. IEEE Trans. Microw. Theory Tech. **MTT-32**(8), 936–937 (1984)
4. P.S. Hall, S.J. Vetterlein, Review of radio frequency beamforming techniques for scanned and multiple beam antennas, in *IEE Proceedings H (Microwaves, Antennas and Propagation)* vol. 137, no. 5 (1990), pp. 293–303
5. Y.J. Cheng, W. Hong, K. Wu, Millimeter-wave substrate integrated waveguide multibeam antenna based on the parabolic reflector principle. IEEE Trans. Antennas Propag. **56**(9), 3055–3058 (2008)
6. N.T. Nguyen, R. Sauleau, M. Ettorre, L. Le Coq, Focal array fed dielectric lenses: an attractive solution for beam reconfiguration at millimeter waves. IEEE Trans. Antennas Propag. **59**(6) PART 2, 2152–2159, (2011)
7. D.M. Pozar, S.D. Targonski, H.D. Syrigos, Design of millimeter wave microstrip reflectarrays. IEEE Trans. Antennas Propag. **45**(2), 287–296 (1997)
8. R.D. Javor, X.-D. Wu, K. Chang, Offset-fed microstrip reflectarray antenna. Electron. Lett. **30** (17), 1363 (1994)
9. S.H. Hsu, C. Han, J. Huang, K. Chang, An offset linear-array-fed Ku/Ka dual-band reflectarray for planet cloud/precipitation radar. IEEE Trans. Antennas Propag. **55**(11) II, 3114–3122 (2007)
10. D.-C. Chang, M.-C. Huang, Microstrip reflectarray antenna with offset feed. Electron. Lett. **28** (16), 1489 (1992)
11. D.M. Pozar, S.D. Targonski, Analysis and design of a microstrip reflectarray using patches of variable size, in *Antennas and Propagation Society International Symposium Digest* (1994), pp. 1820–1823
12. D.M. Pozar, T.A. Metzler, Analysis of a reflectarray antenna using microstrip patches of variable size. Electron. Lett. **29**(8), 657 (1993)
13. C.A. Balanis, *Antenna Theory: Analysis and Design*, 3rd edn. (Wiley, Hoboken, New Jersey, 2005)
14. P. Hannan, Microwave antennas Derived from the Cassegrain telescope. IRE Trans. Antennas Propag. **9**(2) (1961)
15. E.J. Wilkinson, A.J. Appelbaum, Cassegrain systems. IRE Trans. Antennas Propag. **9**(1), 119–120 (1961)

16. W. Hu, M. Arrebola, R. Cahill, J.A. Encinar, V. Fusco, H.S. Gamble, Y. Alvarez, F. Las-Heras, 94 GHz dual-reflector antenna with reflectarray subreflector. IEEE Trans. Antennas Propag. **57**(10), 3043–3050 (2009)

17. D.M. Pozar, Wideband reflectarrays using artificial impedance surfaces. Electron. Lett. **43**(3), 148–149 (2007)

18. S.V. Hum, G. McFeetors, M. Okoniewski, Integrated MEMS reflectarray elements, in *First European Conference on Antennas and Propagation (EuCAP)* (2006), pp. 1–6

19. R. Sorrentino, R.V. Gatti, L. Marcaccioli, Recent advances on millimetre wave reconfigurable reflectarrays, in *2009 3rd European Conference on Antennas and Propagation* (2009), pp. 2527–2531

20. W. Menzel, D. Pilz, M. Al-Tikriti, Millimeter-wave folded reflector antennas with high gain, low loss and low profile. IEEE Antennas Propag. Mag. **44**(3), 24–29 (2002)

21. W. Menzel, M. Al-Tikriti, R. Leberer, A 76 GHz multiple-beam planar reflector antenna, in *32nd European Microwave Conference* (2002), pp. 1–4

22. M.Y. Ismail, R. Cahill, J.A. Encinar, V.F. Fusco, H.S. Gamble, D. Linton, R. Dickie, N. Grant, S.P. Rea, Liquid-crystal-based reflectarray antenna with electronically switchable monopulse patterns. Electron. Lett. **43**(14), 4–5 (2006)

23. W. Hu, M.Y. Ismail, R. Cahill, H.S. Gamble, R. Dickie, V.F. Fusco, D. Linton, S.P. Rea, N. Grant, Tunable liquid crystal reflectarray patch element. Electron. Lett. **42**(9), 509–511 (2006)

24. S.V. Hum, M. Okoniewski, R.J. Davies, Realizing an electronically tunable reflectarray using varactor diode-tuned elements. IEEE Microw. Wirel. Components Lett. **15**(6), 422–424 (2005)

25. P.S. Hall, Progress in Active Integrated Antennas, in *28th European Microwave Conference (EuMC)*, vol. 2, no. 11 (1998), pp. 735–740

26. R.F. Sinclair, E.B. Brown, E.R. Brown, Retrodirective radar for small projectile detection, in *IEEE/MTT-S International Microwave Symposium* (2007), pp. 777–780

27. C.W. Pobanz, T. Itoh, A conformal retrodirective array for radar applications using a heterodyne phased scattering element, in *Proceedings of 1995 IEEE MTT-S International Microwave Symposium* (1995), pp. 905–908

28. E.R. Brown, E.B. Brown, A. Hartenstein, Ku-band retrodirective radar for ballistic projectile detection and tracking, in *IEEE Radar Conference* (2009), pp. 1–4

29. W.P. du Plessis, J.W. Odendaal, J. Joubert, Extended analysis of retrodirective cross-eye jamming. IEEE Trans. Antennas Propag. **57**(9), 2803–2806 (2009)

30. W.P. Du Plessis, J.W. Odendaal, J. Joubert, Experimental simulation of retrodirective cross-eye jamming. IEEE Trans. Aerosp. Electron. Syst. **47**(1), 734–740 (2011)

31. T. Liu, X. Wei, L. Li, Multiple-element retrodirective cross-eye jamming against amplitude-comparison monopulse radar, in *12th International Conference on Signal Processing (ICSP)* (2014), pp. 2135–2140

32. W.A. Shiroma, R.Y. Miyamoto, G.S. Shiroma, J. Tuovinen, W.E. Forsyth, B.T. Murakami, M.A. Tamamoto, A.T. Ohta, Progress in retrodirective arrays for wireless communications, in *Proceedings of the 16th International Symposium on Power Semiconductor Devices & IC's* (2003), pp. 80–81

33. R.Y. Miyamoto, T. Itoh, Retrodirective arrays for wireless communications. IEEE Microw. Mag. **3**(1), 71–79 (2002)

34. E. Sharp, M. Diab, Van Atta reflector array. IRE Trans. Antennas Propag. **8**(4), 1951–1953 (1960)

35. C. Pon, Retrodirective array using the heterodyne technique. IEEE Trans. Antennas Propag. **12**(2), 176–180 (1964)

36. N.B. Buchanan, T. Brabetz, V.F. Fusco, A 62/66 GHz frequency offset retrodirective array, in *IEEE MTT-S International Microwave Symposium Digest, Volume 1* (2002), pp. 315–318

37. K. Chang, Y.-J. Ren, A broadband Van Atta retrodirective array for ka-band applications, in *IEEE Antennas and Propagation Society International Symposium* (2007), pp. 1441–1444

38. A.A.M. Ali, H.B. El-Shaarawy, H. Aubert, Millimeter-wave substrate integrated waveguide passive Van Atta reflector array. IEEE Trans. Antennas Propag. **61**(3), 1465–1470 (2013)

39. S. Christie, R. Cahill, N.B. Buchanan, V.F. Fusco, N. Mitchell, Y.V. Munro, G. Maxwell-Cox, Rotman lens-based retrodirective array. IEEE Trans. Antennas Propag. **60**(3), 1343–1351 (2012)
40. P. Uslenghi, On the generalized Luneberg lenses. IEEE Trans. Antennas Propag. **17**(5), 644–645 (1969)
41. W. Rotman, Analysis of an EHF aplanatic zoned dielectric lens antenna. IEEE Trans. Antennas Propag. **32**(6), 611–617 (1984)
42. A.L. Peebles, Dielectric bifocal lens for multibeam antenna applications. IEEE Trans. Antennas Propag. **36**(5), 599–606 (1988)
43. X. Wu, G.V. Eleftheriades, T.E. Van Deventer-Perkins, Design and characterization of single- and multiple-beam mm-wave circularly polarized substrate lens antennas for wireless communications. IEEE Trans. Microw. Theory Tech. **49**(3), 431–441 (2001)
44. P. Otero, G.V. Eleftheriades, J.R. Mosig, Integrated modified rectangular loop slot antenna on substrate lenses for millimeter- and submillimeter-wave frequencies mixer applications. IEEE Trans. Antennas Propag. **46**(10), 1489–1497 (1998)
45. W. Menzel, A. Moebius, Antenna concepts for millimeter-wave automotive radar sensors. Proc. IEEE **100**(7), 2372–2379 (2012)
46. G. Godi, R. Sauleau, D. Thouroude, Performance of reduced size substrate lens antennas for millimeter-wave communications. IEEE Trans. Antennas Propag. **53**(4), 1278–1286 (2005)
47. N.T. Nguyen, A. Rolland, A.V. Boriskin, G. Valerio, L. Le Coq, R. Sauleau, Size and weight reduction of integrated lens antennas using a cylindrical air cavity. IEEE Trans. Antennas Propag. **60**(12), 5993–5998 (2012)
48. L. Xue, V. Fusco, Patch fed planar dielectric slab extended hemi-elliptical lens antenna. IEEE Trans. Antennas Propag. **56**(3), 661–666 (2008)
49. A. Rolland, R. Sauleau, L. Le Coq, Flat-shaped dielectric lens antenna for 60-GHz applications. IEEE Trans. Antennas Propag. **59**(11), 4041–4048 (2011)
50. J.F. Lotspeich, Explicit general eigenvalue solutions for dielectric slab waveguides. J. Appl. Opt. **14**(2), 327–335 (1975)
51. W. Rotman, R. Turner, Wide-angle microwave lens for line source applications. IEEE Trans. Antennas Propag. **11**(6) 1963
52. W. Rotman, Multiple Beam Radar Antenna System, U.S. Patent No. 3,170,158, 1965
53. E.O. Rausch, A.F. Peterson, W. Wiebach, Electronically scanned millimeter wave antenna using a Rotman lens, in *Proceedings of the 1997 IEEE National Radar Conference*, 1997, no. 449, pp. 374–378
54. E.O. Rausch, A.F. Peterson, W. Wiebach, Millimeter wave Rotman lens, in *Proceedings of the 1997 IEEE National Radar Conference* (1997), pp. 78–81
55. D. Nüßler, Design of a 32 element Rotman lens at 220 GHz with 20 GHz bandwidth, in *German Microwave Conference* (2015), pp. 280–283
56. N. Jastram, D.S. Filipovic, Design of a wideband millimeter wave micromachined Rotman lens. IEEE Trans. Antennas Propag. **63**(6), 2790–2796 (2015)
57. D.S. Filipovic, Z. Popovic, K. Vanhille, M. Lukic, S. Rondineau, M. Buck, G. Potvin, D. Fontaine, C. Nichols, D. Sherrer, S. Zhou, W. Houck, D. Fleming, E. Daniel, W. Wilkins, V. Sokolov, J. Evans, Modeling, design, fabrication, and performance of rectangular M-Coaxial lines and components, in *IEEE MTT-S Inernational Microwave Symposium Digest* (2006), pp. 1393–1396
58. P. Knott, Design of a ridged waveguide feed network for a wideband Rotman lens antenna array, in *IEEE Radar Conference* (2008), pp. 1–4
59. C. Hua, X. Wu, W. Wu, A millimeter-wave cylindrical modified Luneberg lens antenna, in *IEEE MTT-S International Microwave Symposium Digest* (2012), pp. 8–10
60. C. Hua, X. Wu, N. Yang, W. Wu, Air-filled parallel-plate cylindrical modified Luneberg lens antenna for multiple-beam scanning at millimeter-wave frequencies. IEEE Trans. Microw. Theory Tech. **61**(1), 436–443 (2013)
61. O. Lafond, M. Himdi, H. Merlet, P. Lebars, An active reconfigurable antenna at 60 GHz based on plate inhomogeneous lens and feeders. IEEE Trans. Antennas Propag. **61**(4), 1672–1678 (2013)

Chapter 6
Millimeter-Wave Circuits and Components

In practical RF systems, antennas rarely (if ever) operate in isolation. Antennas are usually accompanied by intricate switch networks, power combiners, amplifiers, oscillators, mixers, and high-speed data converters. High frequency circuits such as these have continually enjoyed significant attention from industry and academia, and the development of advanced fabrication technologies, improved semiconductor devices and circuit board materials have profoundly influenced systems for radar, remote sensing, and wireless communication applications.

This chapter will attempt to cover fundamentally important sub-circuits prevalent in all microwave systems (as illustrated in the familiar diagram of Fig. 6.1) as well as discuss the state of millimeter-wave development of these particular circuits. Throughout this chapter we will mention some important parameters that are used to characterize the performance of these circuits, as well as typical architectures employed in millimeter-wave systems to realize the required function.

6.1 High-Speed Data Converters

High-speed and high-resolution analog-to-digital and digital-to-analog converters (ADCs and DACs) are key components in wideband applications and instrumentation systems. Achievable resolution bits and sampling rates have vastly improved in the last decade as a host of new designs surfaced, capitalizing on the ever-increasing sophistication of semiconductor devices, and fabrication processes. One application that can greatly benefit from high-performance ADCs and DACs is software defined radio (SDR). Ideally, an SDR is a radio where the ADCs and DACs are directly connected to the antenna, allowing the device to define and reconfigure all filtering, modulation, and coding blocks in software [1]. Typical SDRs are however, aimed at low cost markets, and as a result, it is generally not

© Springer International Publishing Switzerland 2016
J. du Preez and S. Sinha, *Millimeter-Wave Antennas: Configurations
and Applications*, Signals and Communication Technology,
DOI 10.1007/978-3-319-35068-4_6

Fig. 6.1 Simplified block diagram of a wireless system

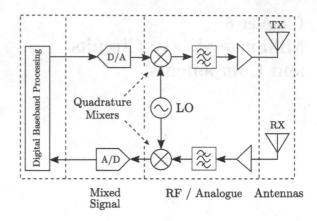

feasible for an SDR to employ a direct conversion approach, although these systems do exist. A more cost-effective alternative involves down-converting the RF signal to an IF, greatly relaxing the performance demands placed on the ADCs and DACs. In spite of the down and up-conversion approach, most SDRs still provide extensive freedom in defining filter and amplifier parameters, on top of offering complete control over the modulation scheme, waveform, center frequency, and signal processing.

6.1.1   ADCs for Millimeter-Wave Systems

The search for high-speed data converters with wide bandwidths and high-resolution capabilities is ongoing, and the performance of these devices must match the ever-increasing data rates required in mobile broadband systems. In 60 GHz systems, ADCs are a crucial component in determining the achievable data rate. One of the associated challenges is related to the power bandwidth product, where the designer needs to balance resolution with sampling rate, and the goal is always to achieve the highest possible sampling rate and resolution without consuming excess power. In their investigation on analog equalization in 60 GHz receivers, Hassan, Rappaport, and Andrews determined that in the presence of multipath and similar effects, a resolution of 5 bits is sufficient to mitigate these effects [2].

An improved resolution/sampling rate combination may be required for some applications in the 60 GHz band (for example, consider the multi-Gb/s media streaming detailed in the IEEE 802.15.3c standard), and Singh et al. [3] recommend at least a 2.5 GS/s sampling rate with 8 effective bits of resolution. Tremendous sampling rates have been shown to be possible with SiGe technology. Shahramian et al. [3] reported a 35 GS/s flash ADC with an effective number of bits (ENOB) equal to 3.7, and an effective 3 dB resolution bandwidth of 8 GHz. The fabricated

ADC used a network of linear buffers to drive a capacitive load connected to a comparator bank. A transimpedance amplifier (TIA) with a 12 dB differential gain was used as a front-end amplifier, and this setup allowed the ADC to process signals in the millivolt range, on top of driving the comparator bank through a symmetric data tree. The ADC developed by Chu et al. [4] reached up to 40 GS/s with a 4-bit resolution, using IBM 8HP SiGe technology. Both these designs show the possibility of direct sampling at millimeter-wave frequencies, although the power consumption in both cases was very high (4.5 W for Shahramian's system, while Chu's system drew 2.1 A from a 3.3 V supply).

With non-CMOS processes, sampling rates in the range of 35–40 GS/s may also be achievable, but such ADCs would only be able to provide very low dynamic ranges unless clock jitter can be effectively controlled [5]. Faced with the high sampling requirements that we have discussed here, flash ADCs are suitable for tackling these challenges and have been the dominant architecture used in high-speed ADCs. While this architecture is quite capable of achieving the highest sampling rate, it suffers from low achievable resolutions as a result of exponentially increasing comparator circuits required as the resolution increases. The metrics that need to be balanced in designing flash ADCs have been described by Uyttenhove and Steyaert [6] and are shown in (6.1)

$$\text{speed} \times \frac{\text{accuracy}^2}{\text{power}} \approx \frac{1}{C_{ox}A_{vt}^2}, \tag{6.1}$$

where C_{ox} is the gate-oxide capacitance and A_{vt} is a process-dependent factor that indicates the mismatch in threshold voltage between subsequent transistors on a die. From the relationship shown in (6.1), we can see that the achievable speed of a flash ADC directly trades with the amount of power consumed in the chip, for a given resolution. This holds for millimeter-wave designs as well, where sub-micrometer processes contribute to faster operation, a lower supply voltage and a greater degree of transistor mismatch. As one could expect, these trends result in greater conversion rates and lower achievable SNR, along with low-voltage architectures being far more prominent. A simplified flash ADC is shown in Fig. 6.2 [7].

A voltage ladder is used to generate a series of voltage references, and these are compared to the input voltage with a number of comparator circuits. The high speed that this design is able to achieve is primarily a result of the absence of a feedback section, which is necessary in many other ADC architectures. Generally, this feedback section consists of a DAC block, notably limiting the achievable throughput. Factors that influence the design of a flash ADC include the encoder type, input capacitance and the number of resolution bits, determined by the number of comparators. The number of comparators grows by $2^n - 1$, where n denotes the number of bits. These parameters balance speed and accuracy, as (6.1) shows. The comparators present a unary code at their output, and this code needs conversion to a standard binary code before it can be processed, which adds to the complexity of the encoder block.

Fig. 6.2 Circuit diagram of a flash ADC

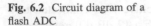
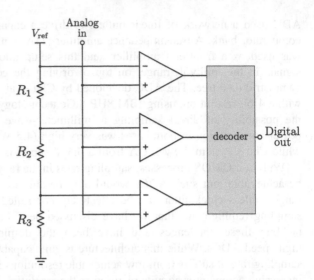

Typically, 6–8 bits are designated as the resolution limit for flash ADCs, and this leads to a fundamental limit on the achievable SNR [7]. The specified resolution requirement trades with power consumption and where error correction is employed, power consumption can also trade with accuracy, provided that error correction is integrated into the ADC design. In general, additional encoding steps or error correction blocks directly affect the maximum speed that the ADC is able to reach.

To reduce the power consumption, several methods have been proposed and evaluated. However, these largely revolve around modifying the architecture to reduce the number of required comparators that have a detrimental effect on the digital output [5]. The process used in construction of the ADC can have a massive effect on its design and performance. The process typically sets the supply voltage, and the design of a flash ADC can be especially challenging when faced with low supply voltages associated with modern CMOS processes.

While flash ADCs are capable of achieving the highest sampling rates, recent results have suggested that other architectures might prove useful for millimeter-wave systems. Ku et al. [8] developed a time-interleaved subranging ADC in a 65 nm CMOS process, which boasts a 40 mW power consumption with a 7 bit resolution and a 2.2 GS/s sampling rate. The time-interleaved design is based on using multiple ADCs in a turn-based sampling scheme. The digital output is derived from each sub-ADC sampling operation, thus allowing theses ADCs to operate at a reduced rate of f_s/M, where M is the number of sub-ADCs used. However, the design of such an ADC is a daunting task, seeing that mismatches in gain, bandwidth, DC offset and clock phase should be minimized. This practically limits the number of sub-ADCs that can be combined in a time-interleaved architecture. Cao et al. also proposed a time-interleaved approach, which used the flash

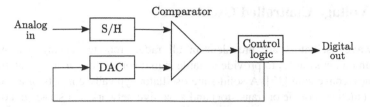

Fig. 6.3 Functional diagram of a successive approximation ADC

Table 6.1 Summary of results for ADC designs reported in the literature

Reference	Topology	ENOB	Power (mW)	Bandwidth (GHz)	Conversion rate (GS/s)	Process
[3]	Flash	3.7	4500	16	35	SiGe BiCMOS
[4]	Time-interleaved	3.5	–	10	40	SiGe BiCMOS
[8]	Time-interleaved	6	40	1.8	2.2	65 nm CMOS
[9]	Time-interleaved	4.5	330	5	10.3	65 nm CMOS
[11]	Pipelined	4.6	9000	6.6	20	180 nm CMOS
[12]	SAR type interleaved	3.5	1200	1.7	24	90 nm CMOS
[13]	Time-interleaved/pipelined	5.1	1600	5	10.3	90 nm CMOS

design as sub-ADCs, and the system achieved 10.3 GS/s conversion rate with a 6 bit resolution [9]. Other groups using this approach have also reported promising results [10–14].

Jiang et al. demonstrated a 1.25 GS/s successive approximation (SAR) ADC, which exhibited power levels as low as 5.28 mW with a 6-bit resolution [15], and this approach does merit consideration for millimeter-wave systems. In a SAR-ADC, a DAC is used in a feedback loop, and its output is compared with an analog input signal. At each iteration of this conversion loop, a single output bit is produced, and the output of the comparator is used to control the DAC output. This is illustrated in Fig. 6.3.

This approach was also used by Louwsma et al. [16] to implement the sub-ADCs required in a time-interleaved design. Ultimately, deciding on a suitable ADC will revolve around balancing power consumption, speed and resolution. Table 6.1 shows a summary of ADCs found in the literature.

6.2 Voltage-Controlled Oscillators

Microwave oscillators are prevalent in all radar, remote sensing and wireless communication systems to provide a stable signal source for carrier generation and frequency conversion [17]. A solid-state oscillator typically consists of a nonlinear device (such as a diode or transistor) and a passive network that serves to convert a DC signal into a sinusoidal RF signal. Employing a frequency multiplier in conjunction with a microwave oscillator is commonly encountered in millimeter-wave systems requiring very high fundamental frequencies.

Despite having lower power handling and frequency capabilities than diodes, transistor-based oscillators present several advantages that justify their usage. Diode devices often pose greater difficulty in integrating with other monolithic integrated circuits, where amplifier and mixer circuits are generally transistor based. Transistor oscillators are very flexible in terms of center frequency, temperature stability, and output noise resulting from the additional freedom available to the designer in choosing the bias point as well as the input and output matching networks. Tunable sources are necessary in many radar and electronic warfare systems, e.g., frequency hopping electronic protection (EP) systems that require rapid alterations of the operating frequency over a wide bandwidth with a dwell time in the order of milliseconds [18]. In 60 GHz systems, large tuning ranges are also required, in order to cover the entire 57–67 GHz band [19], which translates into a range of about 20 % of the center frequency when temperature variation, process inaccuracies and supply voltage variations are accounted for. Other than its tuning range, the performance of a microwave oscillator is determined by measuring frequency stability (in PPM/°C), phase noise (in dBc/Hz at a specified frequency offset relative to the carrier) and the power level of harmonics (in dBc, i.e., relative to the carrier component).

Most oscillators produce a sinusoidal output signal, and this is generally achieved through some form of linear feedback network, as shown in Fig. 6.4.

The output signal V_o can be written in terms of the frequency dependent feedback transfer function $H(\omega)$, and the amplifier voltage gain A, as follows:

$$\frac{V_o}{V_i} = \frac{A}{1 - A \cdot H(\omega)} \tag{6.2}$$

Fig. 6.4 Block diagram of a feedback oscillator

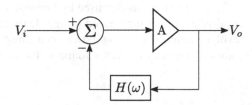

From Eq. (6.2), we can see that there is a particular frequency for which the denominator becomes zero, thus generating an output voltage in the absence of an input voltage, which is the basis of an oscillator. This condition is known as the Barkhausen criterion.

6.2.1 VCO Architectures

Voltage-controlled oscillators (VCOs) can be implemented in a number of different configurations. Some of the most commonly encountered ones are LC-tank VCOs, Colpitts VCOs, distributed VCOs, and sub-harmonic VCOs [20]. A comparison of oscillator configurations fabricated in a SiGe process in the upper millimeter-wave range (220–330 GHz) was recently reported by Tomkins et al. [21]. In an LC-tank VCO, the negative resistance required for oscillation is generated by cross-coupled transistors operating in conjunction with an LC circuit. A key advantage of this configuration is its ability to generate a differential output signal, which is generally preferred in RF circuits, since differential signaling provides improved common mode noise rejection, lower susceptibility to noise and crosstalk and increased robustness against process-voltage-temperature (PVT) variation.

Colpitts VCOs, shown in Fig. 6.5, are transistor based circuits that use LC pi-networks in a feedback loop to generate an oscillation, and their design is quite similar to that of LC-tank oscillators [22].

The Colpitts configuration requires a single transistor, as opposed to the LC-tank circuit requiring two. Furthermore, the Colpitts design is capable of achieving better phase noise performance, but it is slightly more susceptible to low-Q passives as opposed to the tank configuration. Nonetheless, this is a highly popular configuration enjoying widespread implementation, and Winkler et al. [23] was one of the first groups to demonstrate a 60 GHz modified Colpitts oscillator in 2004. A few years later, in 2007, Heydari et al. [24] developed a 104 GHz fundamental mode

Fig. 6.5 Single-ended Colpitts oscillator circuit using a transistor as its amplifier element

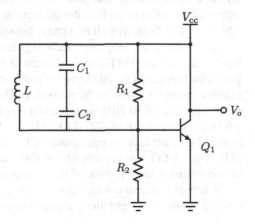

oscillator with a Colpitts topology in a CMOS process. The group determined that, for oscillation to occur at 100 GHz, an inductive load with a Q factor of at least 7.5 is necessary, in conjunction with a shorted CPW transmission line. One of the components in a 165 GHz SiGe transceiver module developed by Laskin et al. [25] was an 80 GHz quadrature Colpitts oscillator with differential 160 GHz output signals. The oscillator was able to achieve a phase noise level of −110 dBc/Hz at a 10 MHz offset from the 84 GHz carrier.

Bipolar SiGe processes have proven to be a popular choice in millimeter-wave designs, and it is no different when concerned with the design and fabrication of VCOs. Nicolson et al. [26] extensively reviewed SiGe BiCMOS VCOs for W-band (77–106 GHz) applications, and developed two prototype designs with MOS and HBT varactors.

Sapone et al. [27] developed a Colpitts-based VCO in a 0.13 μm SiGe BiCMOS process, intended for use in a W-band (particularly at 74 GHz) FMCW radar system. This VCO managed to reach 2 dBm output power from a 2.5 V supply and a phase noise level of −99.3 dBc/Hz at a 1 MHz offset. Li and Rein [28] have demonstrated SiGe VCOs capable of tuning ranges in excess of 25 %, and achieving phase noise levels as low as −110 dBc/Hz at a 1 MHz offset.

In order to increase the power output of an oscillator, a sub-harmonic configuration may be used. This VCO enables active components to operate far below the desired operating frequency, where they produce significantly higher levels of output power than at the desired frequency, which is in contrast with the LC-tank and Colpitts designs which are intended to operate at the fundamental frequency. One subclass of sub-harmonic VCOs is known as push-push VCOs, where the desired frequency component to be extracted is the second harmonic. One disadvantage of conventional sub-harmonic VCOs is that they are only capable of providing a single-ended output, but as Yazdi and Green [29] and Buckwalter et al. [30] have shown, two LC VCOs that generate quadrature clock signals can be used to implement a fully differential architecture. Huang et al. [31] also demonstrated a differential push-push VCO, designed to operate at 131 GHz. The proposed architecture consisted of two pairs of cross-coupled transistors fabricated in a 90 nm CMOS process, and was able to achieve a maximum fundamental power output of 2.1 dBm at each of the differential output ports, a tuning range between 129.8 and 132 GHz, as well as a phase noise of −108.4 dBc/Hz at a 10 MHz offset.

Laskar and Hong employed indium gallium phosphide (InGaP) heterojunction bipolar transistors (HBTs) to implement a 60 GHz push-push VCO [32]. These particular transistors are attractive for millimeter-wave designs because of their low manufacturing cost, reliable fabrication and low flicker noise. The VCO generates a differential pair of 30 GHz signals and a single-ended 60 GHz output, and achieves a 60 GHz phase noise of −77.57 dBc/Hz at a 1 MHz offset. Skafidas and Evans also reported on their take on push-push VCOs, fabricated in a 130 nm CMOS process [33]. Their VCO was capable of a 6 GHz tuning range between 64 and 70 GHz, thus covering a large portion of the unlicensed ISM band at 60 GHz.

Distributed and traveling wave VCOs use a feedback design to generate the desired frequency output along a transmission line. This type of VCO is generally

Table 6.2 Summary of results for VCO designs reported in the literature

Reference	Topology	Phase noise (dBc/Hz)	Output power (dBm)	Bandwidth (GHz)	Center frequency (GHz)	Process
[21]	Push-push	−78 @ 10 MHz	−13.3	21	320	SiGe BiCMOS
[21]	Fundamental	−98 @ 10 MHz	−3.6	27	231	SiGe BiCMOS
[24]	Colpitts	−96 @ 1 MHz	−8.2	3	104	90 nm CMOS
[25]	Differential Colpitts	−110 @ 10 MHz	−3.5	9	84	SiGe BiCMOS
[26]	Differential Colpitts	−101.3 @ 1 MHz	2.5	15	106	SiGe BiCMOS
[27]	Modified Colpitts	−99.3 @ 1 MHz	2	3.6	76	SiGe BiCMOS
[30]	Coupled Subharmonic	−100 @ 1 MHz	−	7	60	SiGe BiCMOS
[29]	Differential push-push	−101 @ 1 MHz	−	−	40	0.18 μm CMOS
[31]	Cross-coupled push-push	−108.4 @ 10 MHz	2.1	2.2	131	90 nm CMOS

capable of a lower end maximum oscillation frequency, in comparison to other VCO configurations, but as Chen et al. have shown, using hybrid design can compensate for this. This VCO, which is described as a triple-push sub-harmonic VCO, was shown to be capable of an extremely wide tuning range, between 100 MHz and 60 GHz [34]. Aside from sub-harmonic VCOs, LO frequencies in the millimeter range can be generated through a frequency multiplier. Where a sub-harmonic VCO generates multiple frequency components and outputs the desired harmonic, the multiplier is dependent on an externally generated signal injected into the oscillator. The multiplier approach has been implemented by several groups, and Räisänen presented an in depth summary of the technology in the early 90s [35], while Zirath et al. [36] conducted a similar study in 2004. Kishimoto et al. [37] developed an injection locked VCO, which relies on multiplying a 15 GHz signal to obtain the desired 60 GHz output signal, therefore providing a multiplication factor of 4.

This approach improves phase noise performance and can be implemented on a very small chip size (2.5 mm × 1.1 mm × 0.15 mm in this case), but it requires a large input power. Furthermore, higher multiplication factors result in a lower achievable output power, which can be disadvantageous in many cases, and the authors stated that temperature variations have a large effect on frequency stability. Table 6.2 summarizes the results discussed in this section.

6.3 Mixers

Mixers are RF and microwave devices used to multiply signals with different frequencies with one another, in the hope of achieving frequency translation [22]. It is a three port device and frequency conversion is achieved through a nonlinear device. With the limited spectrum available today, adjacent allocations are extremely close to one another in frequency. Separation of the desired channel would require extremely high-Q filters, but this problem is somewhat negated by translating the signal frequency to a lower value. Furthermore, multi-bit A/D and D/A converters that operate at bandwidths of several GHz pose several challenges from a design perspective, and in some cases direct conversion is simply not possible. The sampling rate requirement can be significantly reduced if the signal is first mixed down to baseband before it is converted to a digital signal. The same principle holds for the up-conversion path. A functional diagram of a mixer block is shown in Fig. 6.6.

Mixers can be built from either active or passive devices; the former being designed to modulate transconductance, while the latter modulates a switch resistance [38]. Moreover, active mixers use transistors as their amplification elements in order to provide conversion gain, while passive mixers make use of diodes or switching transistors, resulting in a conversion loss. Passive mixers are much simpler to implement than their active counterparts, especially at millimeter and sub-millimeter frequencies. Furthermore, passive mixers are able to achieve very good linearity, possess inherently better noise performance capabilities, and they consume minimal power [39, 40]. At 60 GHz, transistors are operated much closer to their transit frequency, making it increasingly difficult to achieve large LO power levels. This serves to complicate mixer design greatly, and switch-based passive designs suffer the most seeing that they require high power levels to reliably switch.

In order to design adequately functional mixer systems, it is important to balance linearity, conversion loss, power consumption, and isolation. Additionally, since mixers are inherently nonlinear, they will produce intermodulation products. As a result, intermodulation distortion can become problematic, and it is typically represented by the third order intercept point (IIP$_3$).

Minimizing conversion loss can be achieved by properly matching all three mixer ports, although it is treated with greater care in the down-conversion process, since noise figure is a major concern in microwave receivers. This can prove to be a complicated matter however, seeing that several different frequencies as well as their respective harmonics are present.

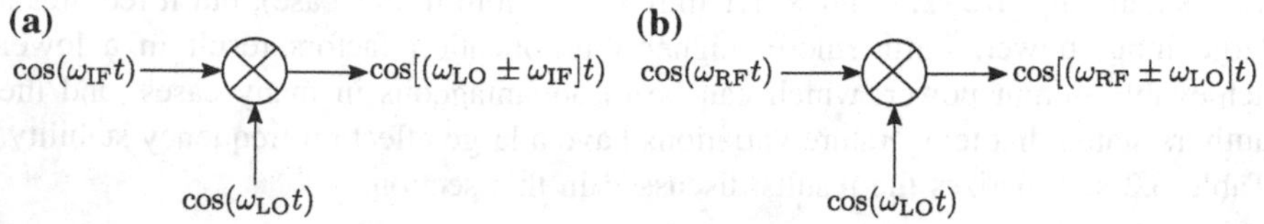

(a)

$\cos(\omega_{IF}t) \longrightarrow \bigotimes \longrightarrow \cos[(\omega_{LO} \pm \omega_{IF}]t)$

$\cos(\omega_{LO}t)$

(b)

$\cos(\omega_{RF}t) \longrightarrow \bigotimes \longrightarrow \cos[(\omega_{RF} \pm \omega_{LO}]t)$

$\cos(\omega_{LO}t)$

Fig. 6.6 Illustration of frequency conversion. **a** Up-conversion. **b** Down-conversion

6.3.1 Mixer Architectures

At a basic level, there are a handful of familiar mixer designs often used in practice. The first of these is a single-ended diode mixer, a passive device, and its circuit diagram is shown in Fig. 6.7 [17].

In the case of a down-conversion mixer shown above, the RF and LO input signals are combined through a diplexer (which can be any microwave device that outputs an in-phase vector sum of the two signals, such as a hybrid junction). The IF, RF, and LO signals in Fig. 6.7 can be rearranged to implement an up-conversion mixer, since the ports are interchangeable. Some designs may require the diode to be biased with a DC voltage, which is decoupled from the RF path with an RF choke. Early millimeter-wave mixer designs made extensive use of diode-based architectures. Oxley et al. presented several single-ended diode mixers at the *IEEE International Microwave Symposium* in 1972, with some designs reaching up to 90 GHz [41]. Siegel and Kerr [42] developed their version of this architecture with a GaAs Schottky diode as the nonlinear element, and the system was designed to operate in the WR-5 waveguide band (140–220 GHz). The authors found that, the effects that result from operating at millimeter-wave frequencies such as charge carrier inertia, plasma resonance, and dielectric relaxation have little effect on the performance of the mixer. Furthermore, the diode mount should be specified up to the fourth harmonic of the LO frequency. Another set of diode mixers was reported by Sisson in the 1980s, with a particular focus on characteristics and performance of newly developed diodes [43]. Following the immense surge in popularity of diode mixers, resistive field-effect transistor (FET) mixers have been shown to provide improved intermodulation distortion and compression performance in comparison. Lin and Ku [44] studied InP-based FET mixers, and compared their theoretical performance to familiar GaAs FET mixers.

The second architecture that we will be covering here is known as a single-balanced mixer, and a circuit diagram of an active, transistor based single-balanced mixer is shown in Fig. 6.8.

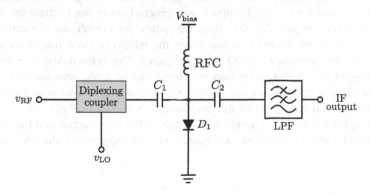

Fig. 6.7 Single-ended diode mixer circuit

Fig. 6.8 Circuit diagram of an active single-balanced mixer

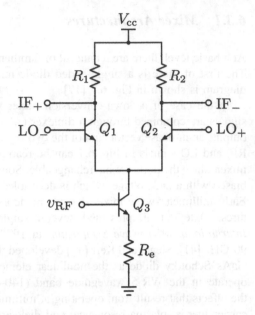

Fig. 6.9 Diagram of a passive single-balanced mixer. Either 90° or 180° hybrid couplers are suitable for this design

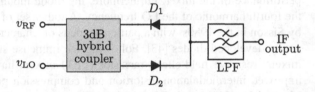

A passive variant of the single-balanced mixer is shown in Fig. 6.9.

Zhang et al. [45] demonstrated the use of substrate integrated waveguide hybrids for single-balanced mixers operating at millimeter-wavelengths.

Another popular class of mixers is the image reject mixer, often encountered in quadrature systems. As we have seen in Fig. 6.6, a multiplying mixer produces both $(\omega_{LO} + \omega_{IF})$ and $(\omega_{LO} - \omega_{IF})$ outputs, and generally only one of these are desired, while the other is regarded as the image frequency. As a result, the unwanted image component must be filtered out, but given the relatively close frequency spacing between the two signals, a high-Q filter is required. This is not a desirable approach, since integrating a high-Q filter into the circuit can be a cumbersome process, and external filtering might prove too costly. The image reject mixer is one solution to this problem, and a diagram of this architecture is shown in Fig. 6.10.

A 0° splitter is used to provide LO signals to the two mixers, and this signal is then mixed with the in-phase and quadrature components of the incoming RF signal.

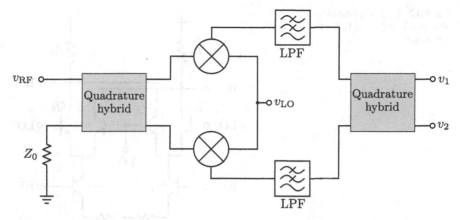

Fig. 6.10 Circuit diagram of an image reject mixer. The output signals v_1 and v_2 denote the upper and lower sidebands, respectively

Alternatively, a quadrature hybrid can be used to split the LO signal, which would require a 0° splitter to separate the RF signal. Gunnarsson, Kuylenstierna, and Zirath have reported on their extended analysis of FET-based image reject mixers [46]. Three mixers were designed for a 1 dB bandwidth of 10 GHz centered at 60 GHz, all having the same topology. A set of single-ended resistive mixers, IF filters, a Wilkinson divider, a branch line coupler, and a quadrature combiner were integrated in all three designs. Similar mixers were used in an earlier design [40]. Double-balanced mixers improve on many shortcomings that are characteristic of single-balanced topologies. Aside from poor matching at the RF input, a double-balanced mixer provides excellent RF to LO isolation and conversion loss, as well as a higher achievable IIP$_3$ out of all the mixers discussed here [17]. A diagram of this topology is shown in Fig. 6.11.

The Gilbert cell mixer (shown in Fig. 6.12) along with its many variants, is by far one of the most popular mixer architectures [22]. It is a natural extension of the

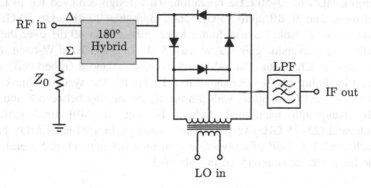

Fig. 6.11 Circuit diagram of a double-balanced mixer

Fig. 6.12 Circuit schematic
of a double-balanced Gilbert
cell mixer

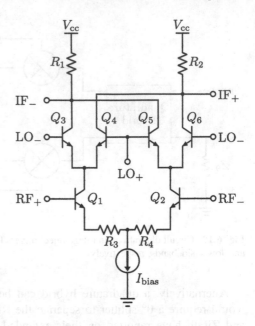

single-balanced mixer, and it requires approximately two times the amount of bias current in order to achieve the same linearity and conversion gain of a single-balanced topology.

Transistors Q_1 and Q_2 are a differential pair that form the driver stage, and the top transistors Q_3 through Q_6 form a switch, driven by the LO. Through symmetry, both the RF and LO signals are cancelled out at the IF output ports, and this also provides high isolation between RF and LO signals. While this topology is compatible with single-ended signals, its full potential can only be reached if differential signaling is used.

The use of Gilbert cell mixers in millimeter-wave designs has been reported on several occasions in the literature. One of the earliest designs was presented by Lin et al. [47] in 2006, where the group developed a Gilbert cell down-conversion stage in 130 nm CMOS for 9–50 GHz operation. This design achieved RF to LO isolation of more than 30 dB up to 30 GHz, and more than 20 dB up to 50 GHz. LO to IF and RF to IF isolation was found to be greater than 40 dB over the entire bandwidth, and conversion gain was at least 5 dB. In a wideband W-band receiver front end-design, Khanpour et al. [48] used a double-balanced Gilbert cell mixer to convert an incoming 74–91 GHz signal to a 1 GHz IF. The system achieved more than 60 dB RF to LO isolation, with a noise figure varying between 8 and 10 dB over the bandwidth mentioned earlier. Tsai et al. [49] implemented an ultra-wideband (25–75 GHz) version of this topology in a 90 nm CMOS process, which achieved 3 ± 2 dB of conversion gain on a 0.3 mm^2 chip. Several mixers from the literature are summarized in Table 6.3.

Table 6.3 Summary of results for mixer designs reported in the literature

Reference	Topology	Conversion gain (dB)	Power (mW)	Bandwidth (GHz)	Isolation (dB)	Process/technology
[42]	Single ended diode	−5.7	–	140–220	–	GaAs
[43]	Single ended diode	−7	–	90–140	–	GaAs
[47]	Gilbert cell	5	97	9–50	40[a], 20[b]	0.13 μm CMOS
[48]	Double balanced Gilbert cell	4	47	75–91	60	65 nm CMOS
[49]	Double balanced Gilbert cell	3 ± 2	93	25–75	30	90 nm CMOS
[50]	Resistive	−12.5	0	51–62	34	65 nm CMOS
[51]	Gilbert micromixer	7	176	79–82	34[a], 28[b]	SiGe HBT

[a]RF-LO
[b]RF-IF

6.4 Power Amplifiers

The power amplification stage is a key component in establishing the link budget and power consumption requirements of a wireless device, and it is traditionally the most power hungry device in the RF chain. As expected, this directly affects a device's battery life, and it is therefore necessary that the amplifier delivers acceptable output power and efficiency while maintaining linearity for a specified modulation scheme. Furthermore, the amplifier gain, stability and power supply requirements are key aspects in assessing the performance of such a device [52]. However, the extremely low voltage of power supplies fabricated in sub-micrometer CMOS processes and the large dynamic range required by modern modulation schemes (one example is orthogonal frequency division multiplexing, or OFDM) make the requirements, listed earlier, difficult to achieve [53]. Dramatic improvements in solid-state technology in the last few decades have led to the vast majority of microwave amplifiers being implemented with transistor devices, such as Si bipolar junction transistors (BJT), GaAs, or SiGe heterojunction bipolar transistors (HBT), GaAs metal–semiconductor field-effect transistors (MESFET), Si metal–oxide–semiconductor FETs, and GaN high electron-mobility transistor (HEMT) [17, 22, 54]. A block diagram describing a basic transistor amplifier is shown in Fig. 6.13, where G indicates the gain in each part of the network, Γ indicates a particular reflection coefficient, and [S] denotes the measured S-parameter matrix for the transistor used [17, 22].

Transistors are suitable for use at frequencies over 100 GHz, and they are suitable for applications that require compact size, wide bandwidth, low noise figure, and reasonably high power capacity.

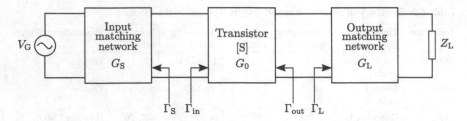

Fig. 6.13 Block diagram of a basic single stage transistor amplifier

6.4.1   Amplifier Figures of Merit

In short, the performance of an amplifier can be assessed through its output power, stability, gain, linearity, efficiency, noise figure, and ruggedness. Noise figure is more of a concern in low-noise amplifiers (LNA), where the gain of the amplifier is intentionally reduced to reach a particular noise figure. To begin, the output power can be determined by measuring the power delivered to a specified load. In terms of amplifier stability, two cases exist, namely, unconditional and conditional stability [22]. Stability is entirely dependent on the S-parameters of the device used, and we can classify an amplifier as unconditionally stable if the following conditions are met for any combination of passive load and source impedances:

$$|\Gamma_{in}| < 1 \tag{6.3}$$

$$|\Gamma_{out}| < 1 \tag{6.4}$$

where Γ once again indicates the reflection coefficient. If the relationships in (6.3) and (6.4) hold for only certain source and load impedance values, we can conclude that the amplifier is potentially unstable, or conditionally stable. Amplifier gain can refer to either voltage gain or power gain, although the latter is encountered more often in practice. Both, however, are heavily dependent on input and output matching networks as well as the amplification element itself. Linearity is determined from analyzing the output power over a range of input power values. Ideally, the amplifier remains linear for any input power, but this is not the case. Instead, a 1 dB compression point is defined as the power level for which the output power has decreased from the ideal linear characteristic by 1 dB, typically denoted as P_{1dB}.

Efficiency provides an indication of the ability of the amplifier to convert DC power into RF power, and it is given by

$$\eta = \frac{P_{out}}{P_{DC}}. \tag{6.5}$$

However, this relationship lacks the inclusion of the RF power delivered to the system, and accounting for this leads us to a quantity known as the power added efficiency (PAE), and this can be computed as

$$\eta_{PAE} = PAE = \frac{(P_{RFout} - P_{RFin})}{P_{DC}}, \tag{6.6}$$

where P_{RFout}, P_{RFin}, and P_{DC} indicate the RF output and input power, and the DC input power, respectively. In a situation where the power amplifier (PA) spends considerable time in each cycle below saturation, the overall efficiency drops, and operating the amplifier at this power level is commonly known as "back-off" [55].

6.4.2 Power Amplifier Architectures

As mentioned, high dynamic range requirements of certain M-ary schemes impose strict linearity requirements on the amplifier, and this is, in part, a result of the peak-to-average power ratio (PAPR) of the scheme. Other than OFDM, single-carrier frequency domain equalization (SC-FDE) is another popular modulation scheme for 60 GHz devices, and it requires a lower PAPR, relaxing the linearity requirement on the PA.

In order to improve the linearity of a PA, several approaches have been suggested. The first of these is the Doherty configuration, where the signal is split between carrier and peaking amplifier paths, each utilizing a different amplifier class, and then combined into one at the output of the network [56]. Other approaches employed include self-biasing [55, 57], envelope tracking [58, 59] and power combining [60]. In some cases where low spectral efficiency is not an issue, an inexpensive device is able to achieve higher efficiencies through using a switching amplifier, as opposed to a linear configuration.

A Doherty architecture combines the outputs of two parallel amplifiers (as shown in Fig. 6.14) in order to improve efficiency. For lower voltage levels at the input, one amplifier operates near saturation (with high efficiency) while the second amplifier is turned off. As the input voltage is increased, the first amplifier continues to operate in saturation, while the second amplifier turns on. Power splitting and combining are the major challenging aspects of this design, together with matching these sections to both amplifiers. Interestingly enough, this problem is accentuated at lower frequencies, mainly due to the required length of transmission lines used in the combiner and splitter networks. As the transmission line sections become shorter, this problem becomes increasingly simple to deal with. The Doherty amplifier developed by Yang et al. [56] exhibited a 14 % PAE at the design frequency of 2.4 GHz.

Fig. 6.14 Block diagram of a
Doherty amplifier

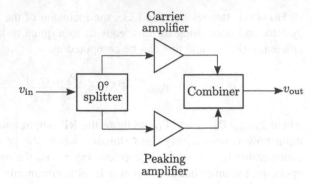

There have been but a handful of reports on Doherty architectures implemented in the millimeter-wave region. Kaymaksut, Zhao, and Reynaert developed one such amplifier in 40 nm CMOS technology, intended for E-band operation at 80 GHz [61]. The proposed architecture combines four push–pull amplifiers by using a 2 × 2 parallel-series transformer combiner network. Using the transformer approach eliminates the need for bulky transmission lines that inhibit the performance of traditional Doherty topologies. The two-stage amplifier achieves an 11.1 % PAE (at the 1 dB compression point) and manages to reach an output power level of 16.2 dBm from a 0.9 V power supply. The P_{1dB} value was found as 15.2 dBm.

Shopov et al. conducted an in depth study on the millimeter-wave extension of the Doherty concept, which covered the effects of breakdown voltage, output impedance and knee voltage in a 90 nm CMOS process [62]. An experimental Doherty PA was designed for 71–76 GHz operation, and an adaptation of the Curtice-Ettenberg model was used to predict design trends and limitations of the active CMOS devices at millimeter-wave frequencies.

A K_a-band (26.4 GHz) Doherty PA was recently reported by Curtis et al. [63] which also utilized a two-stage approach for both the main and auxiliary amplifiers. The amplifier had been fabricated in a 0.15 μm GaAs pseudomorphic high electron mobility transistor (pHEMT) process. At the design frequency, measured results revealed that the PA achieved 25.1 dBm output power at P_{1dB}, as well as a 38 % PAE. The measured small signal gain was 10.3 dB. The same process was used in an earlier design reported by Tsai and Huang [64]. In this case, the amplifier was designed for a 38–46 GHz bandwidth, and achieved a 7 dB small signal gain along with a 21.8 dBm output saturation power.

The second approach to improving linearity is known as envelope tracking, or envelope elimination and restoration (EER). It works by varying the power supply voltage rapidly as the input signal changes, in an attempt to keep the amplifier near saturation. This approach is beneficial when the amplifier is operated below back-off, simply because it allows multiple nonlinear amplifiers to work in conjunction with one another to provide linear amplification. Typically, nonlinear amplifiers are more efficient, because their operating point is closer to saturation

[53]. Choi et al. [59] combined the envelope tracking concept with a Doherty architecture for operation around 1.8 GHz. This amplifier achieved a 38.6 % PAE at an output power level of 24.22 dBm, as well as a gain of 24.62 dB. The measured PAE at 16 dB back-off was 23 %. Although these results are promising, implementing envelope tracking in millimeter-wave amplifiers is not a very practical approach to linearity improvement.

Yan et al. [65] reported a GaAs MMIC-based PA, which was designed for operation around 44 GHz and uses a combination of envelope tracking and digital pre-distortion to improve linearity. The amplifier was tested with 64-QAM signals with a 20 MHz bandwidth and a PAPR as high as 7.6 dB, reaching data rates of up to up to 120 Mb/s. The 7.6 dB PAPR is a result of the time-varying envelope characteristic present in 64-QAM signals, and it poses interesting challenges for millimeter-wave linearity improvement. In the pre-distortion block, the output signal is sampled and converted down to baseband, which is a two-stage process using 44.2 GHz and 1.9 GHz LO signals. The amplifier output is then synchronized in time with the desired signal, allowing the nonlinear behavior of the PA to be extracted. A pre-distorted signal can then be generated by using the inverse of the PA model, and this signal is then sent to the PA to clean the output spectrum and achieve a linearized output. At an output power level of 23.8 dBm, the adjacent channel power ratio (ACPR) was less than –40 dBc, with a 2 % error vector magnitude (EVM). Using an external drain modulator, an improvement in efficiency from 1.22 to 7 % (fivefold increase) was found for the final MMIC stage.

In amplifiers requiring high levels of output power, such as class-E PAs, self-biasing is an attractive approach. Since it regulates the supply and bias voltages, this approach is at least somewhat similar to envelope tracking. Self-biasing regulates these voltages to achieve an optimal distribution of voltage swing for all transistors in the PA, by using a network of passive devices. A properly designed self-biasing network can increase the achievable output power as well as improve the PAE. Apostolidou et al. achieved a 60 % PAE at 30 dBm output power, and a 40 % PAE at 16 dB back-off in their 2 GHz class-E amplifier, using the self-biasing approach [57]. In order to implement this technique at millimeter-wave frequencies, an alternative approach is required to implement the required passive devices. Serhan, Lauga-Larroze, and Fournier recently demonstrated a 60 GHz PA with adaptive biasing, fabricated in 55 nm BiCMOS technology [66]. The amplifier used an on-chip power detector in a control loop in order to dynamically adjust the DC bias current supplied to the PA. The power detector is placed at the output of the amplifier in order to track the output signal envelope. Running off a 1.2 V power supply, this amplifier achieved a peak PAE of 16 %, a 7 dB small signal gain and an output P_{1dB} of 7.5 dBm. In a simulated comparison to the PA without adaptive biasing, an approximate improvement of 18 % in PAE and power consumption was found.

The amplifier architectures that we have discussed here are summarized in Table 6.4.

Table 6.4 Summary of results for power amplifier designs reported in the literature

Reference	Topology	Gain (dB)	Output P$_{1dB}$ (dBm)	Frequency (GHz)	PAE (%) @ P$_{1dB}$	Process
[61]	Doherty	9	15.2	77	11.1	40 nm CMOS
[62]	Doherty	4.7	11.7	60	30.6	90 nm CMOS
[62]	Balanced	7.9	11.9	60	30.1	90 nm CMOS
[63]	Doherty	10.3	25.1	26.4	38	0.15 μm GaAs pHEMT
[64]	Doherty	7	19.6	42	–	0.15 μm GaAs HEMT
[65]	Envelope Tracking	16	32	44	10	GaAs
[66]	Adaptive biasing	7	7.5	60	16	55 nm BiCMOS

6.4.3 Power Combining Techniques

The Doherty architecture is an example of matching sections and power combining being implemented with transmission lines. In fact, power-combining amplifiers have been used for several decades. As early as 1976, a time when newly developed diodes opened up numerous new possibilities in millimeter-wave research, Kuno and English developed a two-stage millimeter-wave IMPATT power amplifier/combiner system [67]. This configuration proved to offer several advantages. First, the limitation in power impedance that exists when multiple devices operate in a single cavity is removed. Hybrid couplers are used between the amplifier cavities, providing high isolation and minimizing potential instability problems that can arise when multiple devices operate in a single cavity. Furthermore, the use of hybrid couplers eliminates the need for ferrite circulators, and much greater bandwidths can be achieved (the experimental model had a 6 GHz bandwidth at a 60 GHz center frequency). Finally, placing each amplifier within its own cavity means that they can be individually tuned, greatly simplifying impedance matching. The system exhibited a 22 dB small signal gain as well as 1 W of CW output power.

Shortly after, Ma and Sun published their Gunn diode combiner network, which exhibited 90 % efficiency at 45 GHz, as well as an output power of 1 W. The circuit consisted of eight Gunn devices, constructed from vapor-phase-grown multiple-epitaxial N-type GaAs [68]. Potoczniak et al. also used Gunn diodes in a novel combiner network, based on dielectric waveguide oscillator circuits [69].

Liu et al. [60] made use of on-chip coupling transformers to implement power combining. In this approach, the system is split into multiple amplifier stages that provide in-phase power amplification. The output voltages at each sub-PA are then combined in-phase to provide a much larger output power level, that is not limited by the restrictions on voltage swing present in each of the smaller amplifiers. Off-chip power combining is also a viable approach, and was used in the mid-1990s by Robertson and Sánchéz-Hernandez [70] in a 60 GHz microstrip patch antenna, integrated with a Gunn diode oscillator. Another group at Rockwell studied spatial power combining at millimeter-waves, utilizing large arrays of solid-state amplifiers [71]. A more recent study of spatial power combining for a number of different antenna and array configurations was done by Emrick and Volakis [72]. Natarajan et al. [73] also used spatial power combining in their IEEE 802.15.3c phased array transmitter.

Depending on the requirements, spatial power combining with an active antenna array or an on-chip transformer-based approach is used in power combining applications. For one, spatial power combining must be able to handle the challenges of on-chip antennas, which are mainly losses that result from wave propagation within the substrate (as opposed to being radiated outward from the antenna aperture). Furthermore, the configuration and circuit placement of phase shifter elements should also be considered. The spatial approach can prove to be extremely useful in reducing the losses associated with on-chip power combining, by using lower loss substrates such as SiGe, and through carefully planning phase shifter networks as to avoid grating lobes [74].

6.5 Closing Remarks

Apart from antenna developments, component technology is crucially important in the continued evolution of millimeter-wave systems. In this chapter we have discussed many of the key building blocks of millimeter-wave systems, in terms of their functionality and performance as indicated by the research community from the research community.

References

1. J. Mitola, *Cognitive Radio—an Integrated Agent Architecture for Software Defined Radio* (Royal Institute of Technology (KTH), 2000)
2. K. Hassan, T.S. Rappaport, J.G. Andrews, Analog equalization for low power 60 GHz receivers in realistic multipath channels, in *IEEE Global Telecommunications Conference (GLOBECOM)* (2010), pp. 1–5
3. S. Shahramian, S.P. Voinigescu, A.C. Carusone, A 35-GS/s, 4-bit flash ADC with active data and clock distribution trees. IEEE J. Solid-State Circ. **44**(6), 1709–1720 (2009)

4. M. Chu, P. Jacob, J.W. Kim, M.R. Leroy, R.P. Kraft, J.F. McDonald, A 40 GS/s time
   interleaved ADC ising SiGe BiCMOS technology. IEEE J. Solid-State Circ. **45**(2), 380–390
   (2010)
5. B. Sivakumar, A.V. Rajaraman, M. Ismail, A 1.33 GSps 5-bit 2 stage pipelined flash analog to
   digital converter for UWB targeting 8 stage time interleaving architecture, in *1st Microsystems
   and Nanoelectronics Research Conference (MNRC)* (2008), pp. 189–192
6. K. Uyttenhove, M.S.J. Steyaert, Speed-power-accuracy tradeoff in high-speed CMOS ADCs.
   IEEE Trans. Circ. Syst. II Analog Digit. Sig. Process. **49**(4), 280–287 (2002)
7. M. Gustavsson, J.J. Wikner, N.N. Tan, *CMOS Converters For Communications* (Kluwer
   Academic Publishers, New York, 2002)
8. I.N. Ku, Z. Xu, Y.C. Kuan, Y.H. Wang, M.C.F. Chang, A 40-mW 7-bit 2.2-GS/s
   time-interleaved subranging CMOS ADC for low-power gigabit wireless communications.
   IEEE J. Solid-State Circ. **47**(8), 1854–1865 (2012)
9. J. Cao, B. Zhang, U. Singh, D. Cui, A. Vasani, A. Garg, W. Zhang, N. Kocaman, D. Pi,
   B. Raghavan, H. Pan, I. Fujimori, A. Momtaz, A 500 mW ADC-based CMOS AFE with
   digital calibration for 10 Gb/s serial links over Kr-backplane and multimode fiber.
   IEEE J. Solid-State Circ. **45**(6), 1172–1185 (2010)
10. S. Gupta, M. Choi, M. Inerfield, J. Wang, A 1GS/s 11b time-interleaved ADC in 0.13 μm
    CMOS, in *IEEE International Solid-State Circuits Conference (ISSCC)* (2006),
    pp. 1931–1936
11. K. Poulton, R. Neff, B. Setterberg, B. Wuppermann, T. Kopley, R. Jewett, J. Pernillo, C. Tan,
    A. Montijo, "A 20 GS/s 8 b ADC with a 1 MB memory in 0.18 μm CMOS, in *IEEE
    International Solid-State Circuits Conference (ISSCC)* (2003), pp. 318–496
12. P. Schvan, J. Bach, C. Falt, P. Flemke, R. Gibbins, Y. Greshishchev, N. Ben-Hamida,
    D. Pollex, J. Sitch, S.C. Wang, J. Wolczanski, A 24GS/S 6b ADC in 90 nm CMOS, in *IEEE
    International Solid-State Circuits Conference (ISSCC)*, vol. 51 (2008), pp. 544–546
13. A. Nazemi, C. Grace, L. Lewyn, B. Kobeissy, O. Agazzi, P. Voois, C. Abidin, G. Eaton, M.
    Kargar, C. Marquez, S. Ramprasad, F. Bollo, V.A. Posse, S. Wang, G. Asmanis, A 10.3GS/s
    6bit (5.1 ENOB at Nyquist) time-interleaved/pipelined ADC using open-loop amplifiers and
    digital calibration in 90 nm CMOS, in *IEEE Symposium on VLSI Circuits* (2008), pp. 18–19
14. O.E. Agazzi, M.R. Hueda, D.E. Crivelli, H.S. Carrer, A. Nazemi, G. Luna, F. Ramos,
    R. Lopez, C. Grace, B. Kobeissy, C. Abidin, M. Kazemi, M. Kargar, C. Marquez,
    S. Ramprasad, F. Bollo, V. Posse, S. Wang, G. Asmanis, G. Eaton, N. Swenson, T. Lindsay,
    P. Voois, A 90 nm CMOS DSP MLSD transceiver with integrated afe for electronic
    dispersion compensation of multimode optical fibers at 10 Gb/s. IEEE J. Solid-State Circ.
    43(12), 2937–2957 (2008)
15. T. Jiang, W. Liu, F.Y. Zhong, C. Zhong, P.Y. Chiang, Single-channel, 1.25-GS/s, 6-bit,
    loop-unrolled asynchronous SAR-ADC in 40 nm-CMOS, in *IEEE Custom Integrated Circuits
    Conference (CICC)* (2010), pp. 1–4
16. S.M. Louwsma, A.J.M. van Tuijl, M. Vertregt, B. Nauta, A 1.35 GS/s, 10 b, 175 mW
    time-interleaved AD converter in 0.13 μm CMOS, IEEE J. Solid-State Circ. **43**(4), 778–785
    (2008)
17. D.M. Pozar, *Microwave Engineering*, 4th edn. (Wiley, Hoboken, New Jersey, 2012)
18. C.D. Schleher, *Electronic Warfare in the Information Age* (Artech House Inc, Norwood,
    Massachusetts, 1999)
19. T. Baykas, C.S. Sum, Z. Lan, J. Wang, M.A. Rahman, H. Harada, S. Kato, IEEE 802.15.3c:
    the first IEEE wireless standard for data rates over 1 Gb/s. IEEE Commun. Mag. **49**(7), 114–
    121 (2011)
20. A.M. Niknejad, H. Hashemi (eds.), Voltage-controlled oscillators and frequency dividers, in
    mm-Wave Silicon Technology: 60 GHz and Beyond (Springer, New York, 2008)
21. A. Tomkins, E. Dacquay, P. Chevalier, J. Hasch, B. Sautreuil, S. Voinigescu, A study of SiGe
    signal sources in the 220–330 GHz range, in *IEEE Bipolar/BiCMOS Circuits and Technology
    Meeting (BCTM)* (2012), pp. 7–10

22. R. Ludwig, B. Gene, *RF Circuit Design: Theory and Applications*, 2nd edn. (Pearson Education Inc, Upper Saddle River, New Jersey, 2009)

23. W. Winkler, J. Borngraber, H. Gustat, F. Korndorfer, 60 GHz transceiver circuits in SiGe:C BiCMOS technology, in *Proceedings of the 30th European Solid-State Circuits Conference* (2004), pp. 83–86

24. B. Heydari, M. Bohsali, E. Adabi, A.M. Niknejad, S. Member, Millimeter-wave devices and circuit blocks up to 104 GHz in 90 nm CMOS. IEEE J. Solid-State Circ. **42**(12), 2893–2903 (2007)

25. E. Laskin, P. Chevalier, A. Chantre, B. Sautreuil, S.P. Voinigescu, 165-GHz transceiver in SiGe technology. IEEE J. Solid-State Circ. **43**(5), 1087–1100 (2008)

26. S.T. Nicolson, K.H.K. Yau, P. Chevalier, A. Chantre, B. Sautreuil, K.W. Tang, S.P. Voinigescu, Design and scaling of W-band SiGe BiCMOS VCOs. IEEE J. Solid-State Circ. **42**(9), 1821–1832 (2007)

27. G. Sapone, E. Ragonese, S. Member, A. Italia, G. Palmisano, A 0.13-μm SiGe BiCMOS colpitts-based VCO for W-band radar transmitters. IEEE Trans. Microw. Theor. Tech. **61**(1), 185–194 (2013)

28. H. Li, H. Rein, Millimeter-wave VCOs with wide tuning range and low phase noise, fully integrated in a SiGe bipolar production technology. IEEE J. Solid-State Circ. **38**(2), 184–191 (2003)

29. A. Yazdi, M.M. Green, A 40 GHz differential push-push VCO in 0.18 CMOS for serial communication. IEEE Microw. Wirel. Compon. Lett. **19**(11), 725–727 (2009)

30. J.F. Buckwalter, A. Babakhani, A. Komijani, A. Hajimiri, An integrated subharmonic coupled-oscillator scheme for a 60-GHz phased-array transmitter. IEEE Trans. Microw. Theor. Tech. **54**(12), 4271–4279 (2006)

31. P.-C. Huang, R.-C. Liu, H. Chang, C. Lin, M. Lei, H. Wang, C.-Y. Su, C.-L. Chang, A.G. P. Vco, A 131 GHz push-push VCO in 90-nm CMOS technology, in *Radio Frequency integrated Circuits (RFIC) Symposium* (2005), pp. 613–616

32. J. Laskar, S. Hong, A 60-GHz push-push InGaP HBT VCO with dynamic frequency divider. IEEE Microw. Wirel. Compon. Lett. **15**(10), 679–681 (2005)

33. E. Skafidas, R.J. Evans, A 60 GHz VCO with 6 GHz tuning range in 130 nm bulk CMOS, in *International Conference on Microwave and Millimeter Wave Technology* (2008), pp. 209–211

34. C. Chen, C. Li, B. Huang, K. Lin, H. Tsao, H. Wang, Ring-based triple-push VCOs with wide continuous tuning ranges. IEEE Trans. Microw. Theor. Tech. **57**(9), 2173–2183 (2009)

35. A.V. Räisänen, Frequency multipliers for millimeter and submillimeter wavelengths. Proc. IEEE **1**(11), 1842–1852 (1992)

36. H. Zirath, T. Masuda, R. Kozhuharov, M. Ferndahl, Development of 60-GHz front-end circuits for a high-data-rate communication system. IEEE J. Solid-State Circ. **39**(10), 1640–1649 (2004)

37. S. Kishimoto, K. Maruhashi, M. Ito, T. Morimoto, Y. Hamada, A 60-GHz-band subharmonically injection locked VCO MMIC operating over wide temperature range, in *IEEE MTT-S International Microwave Symposium Digest* (2005), pp. 5–8

38. A.M. Niknejad, B.A. Floyd, B. Heydari, E. Adabi, B. Afshar, in *Amplifiers and Mixers, in mm-Wave Silicon Technology: 60 GHz and Beyond*, eds. by H. Hashemi, A.M. Niknejad (Springer, New York, 2008)

39. M. Varonen, M. Karkkainen, J. Riska, P. Kangaslahti, K.A.I. Halonen, Resistive HEMT mixers for 60-GHz broad-band telecommunication. IEEE Trans. Microw. Theor. Tech. **53**(4), 1322–1330 (2005)

40. B.M. Motlagh, S.E. Gunnarsson, M. Ferndahl, H. Zirath, Fully integrated 60-GHz single-ended resistive mixer in 90-nm CMOS technology. IEEE Microw. Wirel. Compon. Lett. **16**(1), 25–27 (2006)

41. T.H. Oxley, K. Ming, G.R. Swallow, B.J. Climer, M.J. Sisson, Hybrid microwave integrated circuits for millimeter wavelengths, in *IEEE GMTT International Microwave Symposium* (1972), pp. 224–226

42. P.H. Siegel, A. Kerr, The measured and computed performance of a 140–220 GHz schottky diode mixer, IEEE Trans. Microw. Theory Tech. (MTT) **32**(12), 1579–1590 (1984)
43. M. Sisson, The development of millimetre wave mixer diodes. Radio Electron. Eng. **52**(11), 534–542 (1982)
44. E.W. Lin, W.H. Ku, Device considerations and modeling for the design of an InP-based MODFET millimeter-wave resistive mixer with superior conversion efficiency. IEEE Trans. Microw. Theor. Tech. **43**(8), 1951–1959 (1995)
45. Z.-Y. Zhang, K. Wu, Y.R. Wei, 180-degree substrate integrated waveguide hybrid and its application to broadband millimeter-wave single balanced mixer design, in *Asia-Pacific Microwave Conference* (2010), pp. 1649–1652
46. S.E. Gunnarsson, D. Kuylenstierna, H. Zirath, Analysis and design of millimeter-wave FET-based image reject mixers. IEEE Trans. Microw. Theor. Tech. **55**(10), 2065–2074 (2007)
47. C.S. Lin, P.S. Wu, H.Y. Chang, H. Wang, A 9-50-GHz gilbert-cell down-conversion mixer in 0.13-μm CMOS technology. IEEE Microw. Wirel. Compon. Lett. **16**(5), 293–295 (2006)
48. M. Khanpour, K.W. Tang, P. Garcia, S.P. Voinigescu, A wideband w-band receiver front-end in 65-nm CMOS. IEEE J. Solid-State Circ. **43**(8), 1717–1730 (2008)
49. J.-H. Tsai, P.-S. Wu, C.-S. Lin, T.-W. Huang, J.G.J. Chern, W.-C. Huang, A 25–75 GHz broadband gilbert-cell mixer using 90-nm CMOS technology. IEEE Microw. Wirel. Compon. Lett. **17**(4), 247–249 (2007)
50. M. Varonen, M. Karkkainen, K.A.I. Halonen, V-band balanced resistive mixer in 65-nm CMOS, in *33rd European Solid State Circuits Conference (ESSCIRC)* (2007), pp. 360–363
51. L. Wang, S. Glisic, J. Borngraeber, W. Winkler, J.C. Scheytt, A single-ended fully integrated SiGe 77/79 GHz receiver for automotive radar. IEEE J. Solid-State Circ. **43**(9), 1897–1908 (2008)
52. S.C. Cripps, *RF Power Amplifiers for Wireless Communications*, 2nd edn. (Artech House Inc, Dedham, Massachussets, 2006)
53. T.S. Rappaport, J.N. Murdock, F. Gutierrez, State of the art in 60-GHz integrated circuits and systems for wireless communications. Proc. IEEE **99**(8), 1390–1436 (2011)
54. G. Gonzalez, *Microwave Transistor Amplifiers: Analysis and Design*, 2nd edn. (Prentice Hall, Upper Saddle River, New Jersey, 1996)
55. D.H. Lee, C. Park, J. Han, Y. Kim, S. Hong, C. Lee, S. Member, J. Laskar, A.A. Cmos, A load-shared CMOS power amplifier with efficiency boosting at low power mode for polar transmitters. IEEE Trans. Microw. Theor. Tech. **56**(7), 1565–1574 (2008)
56. L.-Y. Yang, H.-S. Chen, Y.-J.E. Chen, A 2.4 GHz fully integrated cascode-cascade CMOS doherty power amplifier. IEEE Microw. Wirel. Compon. Lett. **18**(3), 197–199 (2008)
57. M. Apostolidou, M.P. Van Der Heijden, D.M.W. Leenaerts, J. Sonsky, A. Heringa, I. Volokhine, A 65 nm CMOS 30 dBm class-E RF power amplifier with 60 % PAE and 40 % PAE at 16 dB back-off. IEEE J. Solid-State Circ. **44**(5), 1372–1379 (2009)
58. P. Asbeck, L. Larson, D. Kimball, S. Pornpromlikit, J.H. Jeong, C. Presti, T.P. Hung, F. Wang, Y. Zhao, Design options for high efficiency linear handset power amplifiers, in *9th Topical Meeting on Silicon Monolithic Integrated Circuits in RF System (SiRF)* (2009), pp. 233–236
59. J. Choi, D. Kang, D. Kim, B. Kim, Optimized envelope tracking operation of doherty power amplifier for high efficiency over an extended dynamic range. IEEE Trans. Microw. Theor. Tech. **57**(6), 1508–1515 (2009)
60. G. Liu, P. Haldi, T.-J.K. Liu, A.M. Niknejad, Fully integrated CMOS power amplifier with efficiency enhancement at power back-off. IEEE J. Solid-State Circ. **43**(3), 600–609 (2008)
61. E. Kaymaksut, D. Zhao, P. Reynaert, E-band transformer-based doherty power amplifier in 40 nm CMOS, in *IEEE Radio Frequency Integrated Circuits (RFIC) Symposium* (2014), pp. 167–170
62. S. Shopov, R.E. Amaya, J.W. M. Rogers, C. Plett, Adapting the doherty amplifier for millimetre-wave CMOS applications, in *IEEE 9th International New Circuits and Systems Conference* (2011), pp. 229–232

63. J. Curtis, A.-V. Pham, M. Chirala, F. Aryanfar, Z. Pi, A Ka-band doherty power amplifier with 25.1 dBm output power, 38 % peak PAE and 27 % back-off PAE, in *IEEE Radio Frequency Integrated Circuits (RFIC) Symposium* (2013), pp. 349–352

64. J.-H. Tsai, T.-W. Huang, A 38–46 GHz MMIC doherty power amplifier using post-distortion linearization. IEEE Microw. Wirel. Compon. Lett. **17**(5), 388–390 (2007)

65. J.J. Yan, C.D. Presti, D.F. Kimball, Y.-P. Hong, C. Hsia, P.M. Asbeck, J. Schellenberg, Efficiency enhancement of mm-wave power amplifiers using envelope tracking. IEEE Microw. Wirel. Compon. Lett. **21**(3), 157–159 (2011)

66. A. Serhan, E. Lauga-Larroze, J.-M. Fournier, Efficiency enhancement using adaptive bias control for 60 GHz power amplifier, in *IEEE 13th International New Circuits and Systems Conference (NEWCAS)* (2015), pp. 1–4

67. H.J. Kuno, D.L. English, Millimeter-wave IMPATT power amplifier/combiner, IEEE Trans. Microw. Theor. Tech. (MTT) **24**(11), 758–767 (1976)

68. Y.-E, Ma, C. Sun, 1-W millimeter-wave gunn diode combiner. IEEE Trans. Microw. Theor. Tech. (MTT) **28**(12), 1460–1463 (1980)

69. J.J. Potoczniak, H. Jacobs, C. M. Lo Casio, G. Novic, Power combiner with gunn diode oscillators. IEEE Trans. Microw. Theor. Tech. (MTT) **30**(5), 724–728 (1982)

70. D. Sanchez-Hernandez, I. Robertson, 60 GHz-band active patch antenna for spatial power combining arrays in European mobile communication systems, in *24th European Microwave Conference* (1994), pp. 1773–1778

71. J.A. Benet, A.R. Perkons, S.H. Wong, A. Zaman, Spatial power combining for millimeterwave solid state amplifiers, in *IEEE MTT-S International Microwave Symposium Digest* (1993), pp. 619–622

72. R.M. Emrick, J.L. Volakis, On chip spatial power combining for short range millimeter-wave systems, in *IEEE Antennas and Propagation Society International Symposium* (2008), pp. 1–4

73. A. Natarajan, S.K. Reynolds, M. Tsai, S.T. Nicolson, J.C. Zhan, D.G. Kam, D. Liu, Y.O. Huang, A. Valdes-Garcia, B.A. Floyd, A fully-integrated 16-element phased-array receiver in SiGe BiCMOS for 60-GHz communications. IEEE J. Solid-State Circ. **46**(5), 1059–1075 (2011)

74. B. Heydari, M. Bohsali, E. Adabi, A.M. Niknejad, A 60 GHz power amplifier in 90 nm CMOS technology, in *IEEE Custom Integrated Circuits Conference (CICC)* (2007), pp. 769–772

Chapter 7
Practical Applications
of Millimeter-Wave Antennas

In the final chapter of this text, we provide a concise summary of millimeter-wave applications that might have appeared throughout the earlier chapters. It is intended as a quick reference, in order to avoid flooding earlier chapters with detailed system-level explanations that derail the discussion from the topic at hand. The interested reader is still encouraged to consult other textbook references that provide much greater detail on some of the topics that we touch on. As we will find, most of the basic concepts that are discussed in this chapter are applicable to a wide range of areas. For example, radars have been used in multiple systems, ranging from vehicle guidance and landing assistance to target detection and tracking. Millimeter-wave radar applied to military systems has been actively researched and implemented since the 1970s. We will briefly discuss some fundamental radar principles and highlight application areas where radar systems have proven to be extremely useful.

Moving away from the military market, millimeter-wave technology in commercial applications has been explored for several decades, highlighted by companies diverting their attention toward commercial products such as automotive radar and millimeter-wave communications [1–4]. These two application areas remain the largest, and greatly beneficial to the industry was the development of sophisticated, low-cost integration processes, and fabrication techniques.

7.1 Communication Systems

While the principles discussed in this section apply to any communication system regardless of application area, our focus will largely remain on broadband mobile communication systems. A block diagram of a basic communication system is shown in Fig. 7.1. Block diagram showing elements of a simple communication

© Springer International Publishing Switzerland 2016   133
J. du Preez and S. Sinha, *Millimeter-Wave Antennas: Configurations
and Applications*, Signals and Communication Technology,
DOI 10.1007/978-3-319-35068-4_7

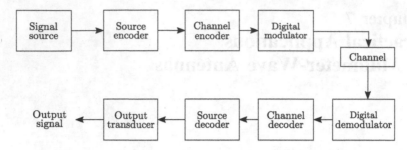

Fig. 7.1 Block diagram showing elements of a simple communication system

system [5]. The signal source may be analog, such as a raw audio or video signal, or a sampled, digital signal. The message signals produced by the source are converted into a coded sequence of binary digits, and ideally the message should be represented by as few bits as possible to minimize redundancy.

The channel encoder is tasked with introducing a signature-type redundancy into the source coded signal, so that the signal can be identified at the receiver. This is also done to reduce the deteriorative effects of the channel. The fully encoded signal is then passed through a digital modulator, acting as an interface to the channel. The primary purpose of this block is to map the digital sequence to waveforms transmitted into free space. For example, if the modulator maps a zero to the signal $s_0(t)$ while mapping a one to the signal $s_1(t)$, the process is called binary modulation. Expanding on this concept, the modulator may be designed to transmit b bits of coded information by using $M = 2^b$ unique waveforms, and this is known as M-ary modulation.

The channel represents a physical transmission medium, which can include copper wire, coaxial cables, free space, and fiber optic cables. The channel invariably introduces some form of corruption into the signal, generally one or more noise-like signals originating from the surrounding environment. The digital receiver is designed to extract the desired signal, which is buried in noise, and reconstruct the desired signal by utilizing the redundancy introduced in the transmitter chain.

7.1.1　Broadband Mobile Systems

Mobile communications are an integral and indispensable part in the lives of an estimated 5 billion people [6], and mobile traffic is experiencing unparalleled growth as a result of the popularity of smartphones and other mobile data devices. Figure 7.2 shows the penetration of mobile broadband in various countries.

Wave propagation at millimeter-wave frequencies is an influential factor when exploring applications that are suitable for millimeter-wave bands. The peak in atmospheric absorption at 60 GHz makes this frequency ideal for short-range

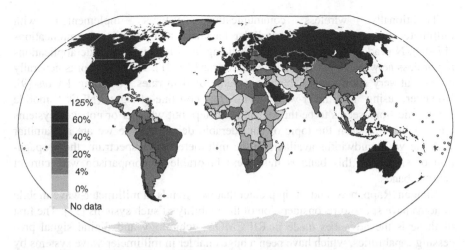

Fig. 7.2 Mobile broadband penetration around the world in 2012, based on the number of subscriptions as a percentage of the population. Image sourced from the International Telecommunications Union (ITU), reprinted here under the Creative Commons Attribution-ShareAlike 3.0 License

wireless links, and since this realization the 60 GHz band has been ruthlessly explored for communication and high-speed media streaming applications [7–9].

In order for the industry to cope with this growth, effective spectrum allocation and improvements in the capacity of mobile systems are critically important. Currently, fourth-generation (4G) systems employ advanced technologies such as multiple-input multiple-output (MIMO) [10–12] and orthogonal frequency division multiplexing (OFDM) to increase spectral efficiency [13, 14]. However, the spectrum between 300 MHz and 3 GHz is becoming increasingly crowded, while the remainder of the 3–300 GHz region is largely unutilized in comparison. Industrial standards supporting multi-Gb/s data rates, such as IEEE 802.15.3c and WirelessHD have been developed as the interest in the unlicensed 60 GHz band grows [9].

Millimeter-Wave Cellular Networks

A mobile network consists of base stations that are geographically situated to optimize coverage, and the actual placement depends on several environmental factors. Realizing the full potential of cellular networks in the millimeter-wave bands does however pose significant challenges, and the feasibility of these networks should carefully be studied and understood. The omnidirectional path loss associated with millimeter-wave transmission can be completely compensated for using directional transmissions and beamforming, but these systems still remain vulnerable to shadowing, which causes inconsistencies in channel quality [15, 16]. Another challenge lies in device power consumption, since a large number of wideband antennas are required.

Traditionally, wireless communication systems implemented with millimeter-waves have been for cellular backhaul and satellite communications [17–19]. Nowadays, the 60 GHz band is being used for high data rate applications in wireless networks and personal area networks [20, 21]. These networks generally operate at very short ranges, but are capable of data rates exceeding 1 Gb/s [9]. However, using millimeter-wave bands for non-line-of-sight (NLOS) mobile communications seems to be the logical next step, but whether or not such systems are feasible, has been the topic of considerable debate. While we are all familiar with the vast bandwidths available in the millimeter-wave spectrum, the propagation of signals in this band is much less favorable in comparison with current cellular bands.

Rangan, Rappaport, and Erkip predict that two trends in millimeter-wave mobile systems have sparked reconsideration of the viability of such systems [15]. The first of these is the advances made in RF CMOS technology and digital signal processing capabilities, which have been a huge enabler in millimeter-wave systems by providing small, low-cost integrated circuits suitable for mobile devices in the commercial sector. Power amplifiers and array combining techniques have made significant progress, and due to the minute wavelengths, large antenna arrays can be fabricated in an area of less than 2 cm^2. This allows multiple arrays to be placed in a single device, providing path diversity.

Second, cellular networks have progressively been evolving toward smaller cells, with support for femtocell and picocell networks (illustrated in Fig. 7.3) that are integral parts of emerging wireless standards [22–25]. In many densely populated urban areas, cells are often 100 m or less in radius, which could place them within range of millimeter-wave signals. Increasing the capacity of existing cells (in terms of users per cell) would be a necessity in the absence of newly allocated spectrum, but it might be a costly affair due to setup and rollout costs. Claussen, Ho and Samuel have estimated that backhaul constitutes 30–50 % of the operating costs involved with cellular networks [26]. In high density areas, millimeter-wave systems may aid cost reduction as a result of their wide bandwidths. This cost reduction effectively provides an alternative method to increase cell capacity.

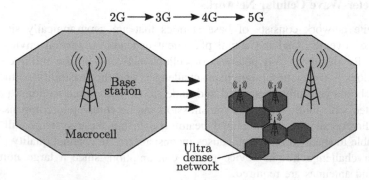

Fig. 7.3 Conceptual diagram of mobile network evolution

Despite the benefits offered by millimeter-wave mobile systems that we have been discussing throughout this text, there are several challenges that need to be intelligently approached if the industry is to advance into the next age of mobile communications. We have mentioned shadowing, and noted that millimeter-wave signals are highly susceptible to this phenomenon. Also referred to as shadow fading, shadowing is the change in attenuation suffered by electromagnetic signals as they propagate through different media. Another type of fading is multipath, where several instances of a signal reach the intended destination at different times as a result of environmental scattering. A brick wall, for example, can attenuate signals in the millimeter-wave range by as much as 40–80 dB [6]. In addition, several groups have studied the effects of the human body in personal area networks and cellular communications, and additional loss of 25–35 dB can be expected [27–29]. However, attenuation due to rain and humidity is not an issue in cellular systems, as opposed to longer range millimeter-wave systems.

The second challenge is associated with range and directional communication. The familiar Friis formula predicts that omnidirectional path loss experienced by a signal in free space is proportional to the square of the system frequency [30]. On the other hand, a smaller system wavelength enables a higher achievable antenna gain for the same physical size. Moreover, recall from our earlier discussion the possibilities of integrating multiple large arrays into a single device, and these arrays are highly directional. This reliance on directional antennas requires a different approach to the omnidirectional approach typically encountered in current communication networks.

For a specific velocity at which a mobile phone travels, channel coherence time is linear in the frequency of the carrier, which means that this will be much smaller in the millimeter-wave range. As an example, a mobile handset traveling at 60 km/h operating on a 60 GHz carrier will experience a Doppler spread of more than 3 kHz, and thus, the channel will vary in the order of several hundred microseconds, significantly faster than current generation cellular systems. High levels of shadowing augment this, and obstacles in the transmission path will create much greater path loss variation [16]. Moreover, millimeter-wave systems will consist of large numbers of picocells and femtocells, which means that relative path loss changes rapidly, and this effect is accentuated by the device switching its associated cell much quicker. This could imply that connection will be intermittent and unstable, and the system should be able to adapt to these changes in order to prevent frequent disconnects.

Another issue lies in coordinating multiple users per link, seeing that current millimeter-wave systems catered toward point-to-point links consisting of a small number of users, where MAC-layer protocols might restrict simultaneous transmissions. New solutions are thus required to adapt millimeter-wave systems to meet specifications of mobile networks.

Massive MIMO

Increasing the capacity of wireless systems through using multiple antennas has been a vigorous field of research for more than 20 years [11]. MIMO installations are commonly encountered in base station installations around the world, but

Fig. 7.4 MIMO system
outline diagram

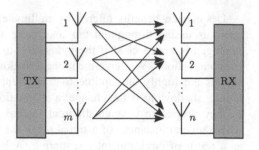

current systems utilize less than 10 antennas, leading to a meek increase in spectral
efficiency. To achieve much greater gains, significantly more antennas need to be
used (orders of magnitude more than the number currently used per base station),
and this concept is often referred to as massive MIMO [6, 31]. In fact, determining
the actual number of antennas required to reach what is thought to be an adequate
improvement in spectral efficiency has been a topic of much discussion [32–34],
and it has been suggested that massive MIMO-like performance can be achieved
with a reasonable, much smaller number of antennas [35]. A simple MIMO system
is shown in Fig. 7.4.

The principle application for massive MIMO is cellular networks, where it is
envisioned that a base station with a large number of antennas is able to provide
simultaneous service to single-antenna users. In a report covering his study of a
base station with the number of antennas approaching infinity, Marzetta predicted
that the effects of uncorrelated noise and fast fading would be eliminated that
spectral efficiency becomes bandwidth-independent, cell size would have no effect
on throughput, and required energy transmitted per bit reduces to nothing [36].
Relatively simple signal processing techniques, such as maximal ratio transmission
(MRT) and combining (MRC), could aid in achieving the advantages of massive
MIMO. While conclusions such as these generally are formed on the assumption of
an ideal propagation environment that exhibits independent Rayleigh fading, where
channels are asymptotically orthogonal and inherently uncorrelated, there is
growing evidence that extremely high spectral efficiencies can be achieved in
nonideal environments with a finite number of antennas.

In a millimeter-wave system, where propagation is generally line-of-sight (LOS),
channel correlation is crucially important. While this would probably prevent the use
of MRT and MRC, methods based on zero-forcing (ZF) or minimum mean squared
error (MMSE) beamforming can possibly be used to combat this factor. Potential
gains in energy efficiency are another key motivator for the massive MIMO concept.
An end user in such a system with idealized channel state information (CSI) can
achieve identical upload throughput as with a single-antenna base station by transmitting only $1/N_t$ of the required power (N_t denotes the number of MIMO antennas)
[37–39]. When an estimate based on MMSE CSI is used, the transmitter power per
user for the same throughput scales with $1/\sqrt{N_t}$, representing a significant reduction
in power consumption for the consumer's device. The same principle holds for
downlink, where each power amplifier (PA) in the system is required to generate no

more than $1/N_t$ of the total output power. These gains in energy efficiency can be exploited to overcome increased path loss, but it also significantly influences design simplicity, efficiency, fabrication cost, and heat dissipation characteristics of the PA.

While millimeter-wave phased arrays have been successfully implemented for a number of application areas, digital beamforming [40] and MIMO through spatial multiplexing [10] has only recently gained interest. A massive MIMO antenna system would have a much smaller footprint than similar designs implemented at microwave frequencies, with the added benefit of significantly more signal bandwidth. At this point, the severity of path loss and environmental attenuation experienced by millimeter-wave signals should be well understood. Fortunately, however, cellular systems operating in this band will focus on small cells in the order of 50–200 m in radius, which means that the effects of 10–20 dB/km attenuation would be minimal. Nonetheless, path loss will still place an upper limit on the cell size, and it can actually be beneficial in scenarios where the cell sizes are exceedingly small, since it allows for greater frequency reuse and limits interference between neighboring cells. On the other hand, the massive MIMO approach could serve to extend range and aid in overcoming path loss.

The received power at the antenna of a communication downlink can be computed by

$$P_r = \frac{G_t G_r}{PL} P_t, \tag{7.1}$$

where G_t and G_r represent the gain of the transmit and receive antennas respectively. In this case, the entire transmit array is taken into account in determining G_t, and it is thus dependent on N_t. For a particular range R, path loss factor n and wavelength, the path loss can be computed as

$$PL = 16\pi^2 \left(\frac{R}{\lambda}\right)^n. \tag{7.2}$$

Wave propagation in urban environments has been studied, among others, by Rappaport et al. [16]. In LOS tests, path loss exponents of 1.7–2.5 were found, and this increased to 3.5–6 in non-LOS tests. Variables that the designer can control, to a degree, that can mitigate the effects of path loss are large array gains, improved receiver noise figure and sensitivity, and transmit power levels. With the current state of millimeter-wave transceivers, it appears that sufficient gain is available, but the number of users that can be serviced in various conditions still needs to be determined. Additionally, millimeter-wave signals are severely attenuated when propagating through brick and concrete structures, which could mean that such cells can only be either indoors or outdoors, limiting their application.

Signal processing as well as circuit and communication system design are the main aspects of millimeter-wave MIMO systems that require careful consideration in the implementation process. Delay spread experienced in millimeter-wave systems can be greatly reduced by using a narrow beam, produced by a MIMO array, in turn reducing the requirement for equalization. On the other hand, very large bandwidths per user would lead to frequency selective fading, for which equalization or modulation can be considered. As we have talked about in prior sections, power efficiency is a key advantage offered by massive MIMO systems. Very high peak-to-average power (PAPR) ratios are characteristic of OFDM schemes, counteracts this advantage, and could serve to restrict downstream performance.

Pitarokoilis, Mohammed, and Larsson determined that single-carrier modulation (SCM) schemes can achieve sum-rate performance close to their theoretical maximum in massive MIMO systems [41]. This is based on systems operating at low ratios of transmit power versus receiver noise power, with an equalization-free receiver and completely independent of the power delay profile unique to the channel. Furthermore, the results are based on independent Rayleigh fading channels, which do not hold in the millimeter-wave bands. Seeing that SCM schemes can be designed with greatly improved PAPR performance, this approach could still prove interesting for power efficiency concerns.

OFDM employed in 4G systems exploits simplified equalization and flexible resource allocation, by orthogonally multiplexing multiple narrowband subchannels in frequency [42]. An alternative to the classical OFDM scheme is known as the constant-envelope (CE) OFDM concept, designed to eliminate the PAPR issue that is consistent with OFDM systems [43]. In order to simplify equalization, a cyclic prefix transmission is used, similar to the approach in conventional OFDM. An important result from the work done by Thompson et al. is that in accounting for the effects of nonlinear power amplification, CE-OFDM performs better than conventional in multipath fading channels. Furthermore, it has been shown that CE-OFDM is capable of achieving better bit error rate (BER) performance as well as better fractional out-of-band power. A key challenge in the design of such a system however, is to avoid the FM thresholding effect that can occur in the phase demodulation and phase unwrapping blocks. Two possible solutions that have been suggested oversampling and phase-locked loops. Nevertheless, by now it should be quite clear that the optimal modulation scheme for millimeter-wave massive MIMO systems is dependent on a multitude of factors and requires significant research.

7.2 Radar

Radar systems are capable of detecting targets that dwell within their search volume, as well as measure key parameters involving that target: range, velocity and angle. At its core, a radar system consists of a transmitter, receiver, antenna and a transmit/receive (TR) switch and of course a target,[1] as shown in Fig. 7.5 [44–47]. The transmitter typically requires a waveform generator to generate an appropriate

---

[1]The target aircraft in Fig. 7.5 was obtained without making alterations to the original content from SudsySutherland (http://sudsysutherland.deviantart.com/), and it is reused here under the Creative Commons Attribution-ShareAlike 3.0 License.

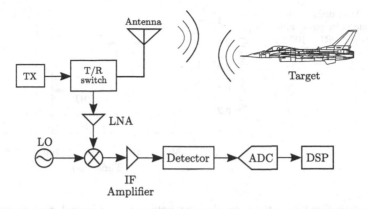

Fig. 7.5 Block diagram of a mono-static radar system [45]

baseband signal that is later used in the detection process, and a typical up-conversion chain that includes filtering and amplification. On the receiver end, a typical down-conversion chain is used to provide in-phase and quadrature (IQ) data to a radar signal processor.

For a given target radar cross section (RCS, denoted by σ), the power received by the radar receiver (P_R) can be determined by the radar range equation,

$$P_R = \frac{P_t G_t G_r \lambda^2}{(4\pi)^3 R^4 L_s} \sigma, \tag{7.3}$$

where

P_t is the transmitted power, in Watts
G_t and G_r indicate transmitter and receiver gain in decibels,[2] respectively
λ is the system wavelength in free space, in meters
L_s is the sum of losses in the system, in decibels
R is the radial distance to the target, in meters.

RCS is a function of several target characteristics, radar parameters as well as the radar-target geometry [45]. Specifically, target RCS is dependent on target material composition and geometry, radar wavelength, transmitter and receiver polarization, and relative positions of the radar transmitter and receiver. As the wavelength is decreased to be much smaller than the size of the target, a phenomenon called high-frequency scattering occurs. The target no longer exists as a single scatterer, but rather consists of multiple scattering centers from which localized energy is reflected. The practical consequence of this is that higher frequency radars can be used to generate extremely accurate RCS measurements of the target and provide a high-resolution target profile.

---

[2]Decibel values must be converted to their linear equivalent before attempting to evaluate (7.3).

Fig. 7.6 Time domain plot
of a simple radar pulse with
$\tau = 20$ µs and PRI $= 100$ µs

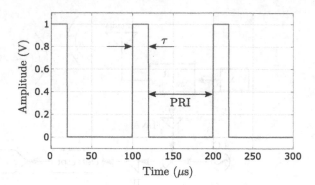

Operational radar systems are divided between pulsed and continuous wave
(CW) operation. In a pulsed radar, several pulses are transmitted coherently
(although a handful pulsed radars operate incoherently), spaced in time by the pulse
repetition interval (PRI). An example of such a pulse train is shown in Fig. 7.6.
Based on the internal timing configuration in the radar, the receiver then integrates
(coherently or noncoherently) the individual return signals from each of these
pulses (since the system has complete control over TX/RX timing) to in order to
improve the SNR of the system. Target range estimation is done by computing the
time delay of a received echo that has passed the radar's detection algorithm. In
other words, the received signal at a particular point in time (or over a range of
samples) crosses the detection threshold, and the radar signal processor decides that
a target is present at that time.

The achievable range resolution is determined by the width of the transmitted
pulse, and it is given by

$$R = \frac{c\Delta\tau}{2},\tag{7.4}$$

where R indicates the radial distance to the target from the radar antenna, c is the
free space propagation constant, and $\Delta\tau$ indicates the pulse width [44–47]. This
relationship can also be written in terms of bandwidth, since the bandwidth B of a
τ-second pulse is equal to $1/\tau$. Furthermore, the estimated target range can be
computed from (7.4) by replacing $\Delta\tau$ with the round-trip time. One advantage of
millimeter-wave frequencies in radar should be apparent at this point, and that is the
large increase in available bandwidth, which in turn enables much greater resolution
in radar measurements.

Pulsed radars are capable of measuring relative velocity through exploiting the
Doppler shift, and multiple pulses are used in order to increase the Doppler resolution and simplify the estimation process. When the target is moving in a radial
direction relative to the radar antenna, the reflected signal will experience a frequency shift that is dependent on the system wavelength and the relative velocity.
The radar can then perform Doppler processing on the received signal to determine

in which resolution bin the target return resides in. Once this is known, an estimate of the target velocity can be made.

The bandwidth and resolution of the Doppler spectrum are determined by the number of pulses used in a transmit burst, and the pulse repetition frequency (inverse of the PRI, denoted by PRF). The PRF is divided into N resolution bins, where N indicates the number of pulses used in the transmit burst. The frequency is then estimated first by computing a $M \geq N$-point discrete Fourier transform (DFT) and then interpolating between DFT bins.

As opposed to the familiar pulsed configuration, a CW radar transmits and receives simultaneously until the system is deactivated. A popular method of determining the range of a target is to use a frequency modulated CW (FMCW) waveform, which effectively puts a timing mark on the waveform. This timing mark can be processed at the receiver to determine the target range. In the absence of a Doppler shift, the difference frequency (sometimes referred to as the beat note, denoted by f_b) is a measure of target range and $f_b = f_r$, where f_r is the frequency deviation that results solely from the target being at range. The beat frequency can then be determined by

$$f_r = \dot{f}_0 \cdot \frac{2R}{c}, \tag{7.5}$$

where $\dot{f}_0$ indicates the modulation rate of the carrier signal [48]. In practice, however, the waveform frequency cannot be continuously changed in one direction, and periodicity is introduced in the waveform. With this in mind, the beat frequency can then be found as

$$f_r = \frac{4f_m R \Delta f}{c}, \tag{7.6}$$

where f_m indicates the modulation rate and Δf is the range over which this modulation occurs. Therefore, we can determine the range of a target by measuring the beat frequency f_r. Including the Doppler shift $f_d = (2v/\lambda)\cos\theta$ to account for a target moving at velocity v, (7.6) can be modified to include this term, enabling the radar to measure velocity. Given the range-Doppler coupling inherent to the FMCW waveform, multiple sweeps with different ramp profiles have to be performed if it is necessary to measure range and velocity for multiple targets simultaneously. An example of a FMCW waveform is shown in Fig. 7.7.

Many radar processing techniques are associated with increasing SNR, which is directly related to the probability of detection (P_D) [44–47]. One popular technique is known as pulse compression, where the bandwidth of the radar pulse is increased to achieve better range resolution with the same pulse width [49]. Furthermore, using this technique allows the designer to increase the pulse width, thereby increasing the average power (P_{avg}) per pulse and improving detection performance. Pulse compression effectively decouples resolution and energy, at the cost of additional processing at the receiver.

Fig. 7.7 Real part of an FMCW waveform with an upwards sweep

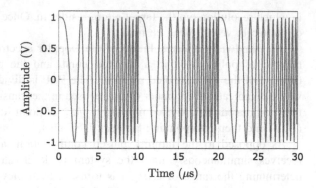

In order to understand factors that influence target detection, the radar equation given in (7.3) is a sensible starting point, and it can be written in terms of the SNR [45] as

$$\text{SNR} = \left(\frac{P_{\text{avg}}T_d}{N\tau}\right) \frac{G_t G_r \lambda^2 \sigma n_p}{(4\pi)^3 R^4 k T_0 \text{FBL}_s} \tau\beta. \tag{7.7}$$

where

P_{avg} is the average pulse power, in Watts
n_p is the number of pulses used in the integration process
τ is the width of the radar pulse, in seconds
T_0 is the standard temperature, 290 K
F is the receiver noise figure, in decibels
β is the pulse modulation bandwidth, in Hertz
N is the processing gain resulting from coherent integration, in decibels
B is the bandwidth of the unmodulated pulse, in Hertz
k is Boltzmann's constant, 1.38×10^{-23} Ws/K
T_d is the Dwell time on the target, in seconds.

For a given beamwidth specification, antenna apertures reduce in size as the frequency of operation increases [50]. Therefore, millimeter-wave radar sensors can provide narrower antenna beamwidths as well as larger absolute bandwidths for the same percent-bandwidth value, both of which serve to increase the attainable SNR. On the other hand, millimeter-wave antennas are generally capable of producing lower gain values, but this is an acceptable tradeoff since the range at which these radars operate are much less than conventional systems. An illustration of resolution in range and angle is shown in Fig. 7.8.

Range resolution for a pulsed radar was given in (7.4), and the equation remains similar for the FMCW case, except that pulse width τ is replaced with the sweep bandwidth $1/\Delta f$. Furthermore, velocity resolution in a FMCW radar can be computed as

Fig. 7.8 Illustration of range and angular resolution in a radar system

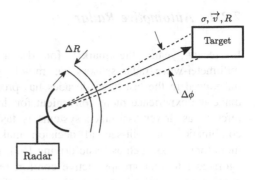

$$\Delta v = \frac{\lambda}{2T}. \tag{7.8}$$

Angular resolution in a radar system is primarily dependent on the antenna system, and the approach differs between search and track modes. In search radar, the system is only interested in detecting a target, while tracking radar is tasked with keeping the antenna boresight pointed at the target, within a certain resolution specification. Tracking radars almost universally employ a technique known as monopulse, where multiple antennas are used to derive a correction angle from the received signal [51].

7.2.1 Civil Applications of Millimeter-Wave Radar

Development on millimeter-wave radars was initially focused on military applications, and the cost of systems and components in the 1970s meant that exploring civil applications would be infeasible [52]. Several important developments led to mainstream implementation of millimeter-wave systems, and some of these developments are transistors with cutoff frequencies that exceed 100 GHz, automatic assembly of planar circuits, reliable, and low-cost MMICs for millimeter-waves, and multilayer, multifunctional circuits.

Clark and Durrant-Whyte have demonstrated the use of a 77 GHz radar as a guidance sensor for autonomous navigation in land vehicles, using an extended Kalman filter to optimally combine range and angle measurements with a vehicle control system [53]. An improved version of this system was later developed that enabled identification of natural features in the environment by using the reflected polarization [54].

The use of synthetic aperture radar (SAR) has been investigated for aircraft landing assistance systems, and Sadjadi et al. have reported such a system [55]. The so-called synthetic vision system consisted of a scanning radar operating at 35 GHz, a heads-up display (HUD) and real-time digital signal processing hardware. Jain has reported on the use of millimeter-wave radars in airport surface surveillance systems [56].

7.2.2   Automotive Radar

In recent years, the market for driver assistance systems that rely on millimeter-wave radar sensors has rapidly grown. The new generation of cars introduced in the last three decades has promised to improve driving safety and make the experience more convenient for drivers. The functionality that we now refer to as driver assistance systems is tasked with relieving drivers from the combination of split-second decision making within complex scenarios and monotonic tasks such as basic driving [57]. Driver assistance systems can be categorized into four groups: active comfort (e.g., adaptive cruise control), passive comfort (e.g., parking assistance), active safety (e.g., automatic braking) and passive safety (e.g., airbags). While passive systems only react to certain scenarios and are not able to influence the vehicle motion, active systems can indeed influence vehicle dynamics such as breaking and accelerating. In order to realize these concepts and implement them in a practical system, a range of different radar sensors is required to make the vehicle aware of its surroundings.

Frequency Allocation and Regulations

Given that radar systems can measure radial distance, velocity and in some systems, angle, they are a key component in driver assistance systems. Furthermore, radars generally are quite robust in adverse weather conditions and bad lighting, making them all the more attractive for these systems. Systems that use millimeter-wave radar technology first appeared in the 1970s [57]. The International Telecommunications Union's (ITU) decision at the 1979 World Administrative Radio Conference (WARC) to support sensor applications in millimeter-wave bands above 40 GHz has been one of the key driving factors [58].

Clutter returns from ground reflections, buildings, adjacent vehicles and guard rails constitute a large portion of the return signal captured by an automotive radar sensor. While sophisticated signal processing can serve to reduce the effect of the clutter response, or separate desired targets from clutter, using higher frequencies can also be beneficial. The ability of an antenna to produce extremely narrow beams at millimeter-wavelengths can provide effective spatial filtering against background clutter, and greatly increase the angular discrimination ability of the radar. Moreover, since automotive radars operate at very short ranges, it would seem that operating in millimeter-wave bands would yield optimal performance.

Two frequency bands primarily used in automotive radar are centered at 77 and 24 GHz. Frequencies outside of this range, i.e., below 10 GHz and above 100 GHz have been explored, but with little practical relevance. The 77 GHz band offers greater possibilities to implement high-performance sensors, but it involves greater difficulty in system design from an engineering perspective. Nonetheless, there are several factors that motivate the continued use of this frequency band. The size of a radar sensor is determined by the antenna aperture. Operating the radar at 77 GHz allows for a physically small antenna capable of achieving a very narrow beam-width. Using the Rayleigh criterion to determine angular resolution, Hasch et al. demonstrated that for the same antenna aperture, a resolution of 5.4° is obtainable at 77 GHz, while only 17.5° is possible at 24 GHz [57]. Reducing the size of the sensor also aids integration challenges, since it allows the design to contend with greater size and weight constraints. Significant improvements in fabrication technology in the last two decades have brought the manufacturing costs of 24 and 77 GHz systems relatively close to one another.

Improving range resolution almost invariably requires larger bandwidth wave-forms, and in terms of percent-bandwidth, the 77 GHz system is the better option. Furthermore, emission regulations in most regions do not permit high power (greater than −40 dBm), high-bandwidth (larger than 250–300 MHz) radiation at 24 GHz, while this combination is not restricted at 77 GHz. Behind-bumper inte-gration is the only concern for sensor operation in the 77 GHz band, where high-permittivity metallic based paints could cause significant reflections at the radome. Pfeiffer and Biebl have suggested a narrowband solution using inductive strips and demonstrated the effectiveness of this technique in [59].

Mercedes-Benz was the first automobile manufacturer to introduce radar-based autonomous cruise control (ACC) in 1999 [60]. Since then, 77 GHz radar sensors have been used in collision mitigation and pre-crash sensing systems, among others. Newer generations of sensors will improve field of view, range, and angular res-olution, as well as minimum and maximum range. Nonetheless, aside from tech-nical advancements, frequency regulation will continue to play a pivotal role in the development of millimeter-wave radars. Difficulties arise from the fact that coun-tries generally have their own regulations regarding spectrum usage, and in rare cases these allocations conflict with neighboring countries. In the millimeter-wave region, the two major allocations are the 76–77 and 77–81 GHz bands, the latter having replaced ultra-wideband (UWB) automotive radars in the 24 GHz band in Europe [57]. Moreover, the 76–77 GHz band permits higher maximum transmitter power levels (55 dBm EIRP[3]), while the 77–81 GHz band provides increased bandwidth and lower permitted power levels [61]. The maximum allowable power spectral density in the latter band is specified to be lower than −3 dBm/MHz, along with a peak limit of 55 dBm EIRP [61]. Additionally, the mean power density outside a vehicle is limited to −9 dBm/MHz, which accounts for the attenuation that result from installing a sensor behind a painted bumper. In the US, emission limits vary based on several vehicle parameters, but it is expected that the regulations will soon adapt to similar values used in Europe.

---

[3]Effective isotropically radiated power.

Classification of Automotive Radars

There are three principle groups of automotive radar, distinguished by the range at which they are intended to measure target parameters:

* Short Range Radar (SRR)—sensing in close proximity of the vehicle, such as parking assistance and obstacle detection systems.
* Medium Range Radar (MRR)—sensing at medium distance and with an average speed profile, such as cross-traffic alerting systems.
* Long Range Radar (LRR)—sensing at longer distance where a narrow antenna beam is required in a forward-looking direction, such as an ACC system.

As we've discussed earlier, resolution is an indication of the ability of the radar to distinguish between targets, whether it be in range, angle, or Doppler shift (radial velocity). On top of the achievable resolution, there still exists a level of uncertainty in the accuracy of the measurement, which is generally much smaller than the resolution itself. A key challenge in the design of automotive radar systems is the separation of closely spaced targets with vastly different RCS values (such as a bus and a motorbike), that may be traveling at the same distance and velocity relative to the radar. This separation can be achieved by designing the radar to have high dynamic range and small resolution in any one of the measured parameters—range, velocity, or angle.

Recent Developments

Before the commercial release of automotive radars in 1999, the concept had been explored and developed around the world [3, 58, 62–64]. With systems becoming increasingly sophisticated, computing power rapidly advancing, and continuous advances in component technology, the next generation of vehicle safety systems is becoming more and more prevalent. Soon after the turn of the century, a group from Infineon Technologies began investigating SiGe bipolar technology for use in automotive radar [65]. The group developed building blocks of millimeter-wave radars in the 77 GHz band (such as VCOs and mixers) and achieved a transit frequency of 200 GHz, a maximum oscillation frequency of 275 GHz, and a ring oscillator gate delay of 3.5 ps.

Gresham et al. [66] were one of the first groups to publish on commercial automotive radar modules. The project was focused on cost reduction and manufacturing simplicity, and the circuit was implemented using discrete GaAs/AlGaAs pHEMTs, GaAs Schottky and varactor diodes, as well as pHEMT MMICs installed on a 127 μm glass substrate. Lutz et al. [67] investigated the use of compressive sensing in automotive radars, along with experimenting with fast chirp modulations. Sawade et al. [68] demonstrated a cooperative active blind spot assistant, and Dudek et al. [69] recently reported a detailed investigation of adaptive beamforming in phased arrays, with a focus on automotive safety. The 77 GHz system utilized a novel antenna control method, which adaptively couples the radar sensor to the vehicle steering angle, thereby directing the beam together with the ego-vehicle into the curve. This has proven to greatly increase measurement range and thus provides additional time for the safety system to react.

7.3 Imaging

Heightened levels of security at different checkpoints (such as airports and train stations) have become increasingly important. Familiar systems such as metal detectors and hand-luggage X-rays have been used for many years to great effect, but suffer from shortcomings that need to be addressed for future systems. For example, metal detectors can only detect metallic objects, such as handguns and knives, and the detection probability relies on a number of factors that cannot necessarily be accounted for in the design of such a system. Furthermore, since these systems are only able to detect these items, and are not able to distinguish between harmless items, such as glasses, keys and belt buckles, and items that could pose a security threat [70]. Two commercial imaging systems are shown in Fig. 7.9.

It is clear that newly developed systems, such as ones based on high-resolution imaging, could be a solution to this problem. While X-ray imaging systems have been proven to be effective, perceived negative health effects might limit mainstream acceptance of such systems. On the contrary, millimeter-waves are non-ionizing, and therefore pose no health hazard even at moderate power levels. Imaging systems that operate at millimeter-wavelengths can penetrate clothing to form an extremely high-resolution image of a person and any concealed items around the body. Yujiri and Shoucri [71] have discussed passively using millimeter-waves as a viable technique for several imaging applications. Newly developed sensors for millimeter-wave systems have renewed interest in this field and enabled generation of images at extremely high data rates.

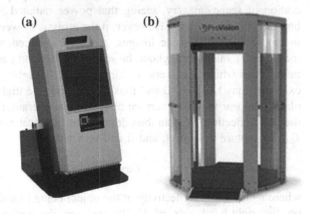

Fig. 7.9 **a** Passive and **b** active commercial millimeter-wave imaging systems

A major benefit of these types of imaging systems is the ability to create clear images in just about any type of weather condition; smoke, sandstorms, clouds, and fog. One advantage that a passive system brings to the table, is the ability to operate covertly (that is, without emitting any form of radiation, unlike radar, and lidar systems), similar to infrared (IR) and visible sensors. Furthermore, clutter variation is much lower in passive imaging systems and the signature of metallic objects can also be clearly defined. These attributes serve to greatly reduce the false alarm rate in automated detection schemes. Military applications that could benefit from this include surveillance, refueling in low-visibility conditions, search and rescue, and precision targeting. Civilian systems could similarly benefit from this technology, such as aircraft landing, harbor surveillance, traffic monitoring, and concealed weapon and contraband detection [72–75].

7.3.1 Millimeter-Wave Radiometry

In the millimeter-wave bands, objects reflect and emit radiation, just as they would in the visible and IR domains. The degree to which this reflection (or emission) occurs is determined by the polarization-dependent emissivity of the object, denoted by ε [71]. A perfect radiator is called a blackbody, and it has an emissivity $\varepsilon = 1$. On the other hand, a perfect reflector has $\varepsilon = 0$. The emissivity of an object is primarily dependent on surface roughness, the angle of observation, and the dielectric properties of the constituting materials. The surface brightness temperature T_s, also known as radiometric temperature, of a particular object, can be determined as

$$T_s = \varepsilon T_0, \qquad (7.9)$$

where T_0 denotes the physical thermodynamic temperature unique to the object in question. Emissivity is an important quantity that plays a pivotal role in the generation of scene imagery, seeing that power radiated from different objects vary based on their emissivity. However, if the emissivity were the only influential factor in the formation of scene images, the actual procedure would involve mapping measured T_s values throughout the scene. The second part of the equation lies in the method by which the scene is illuminated. A highly reflective metal plate, for example, may have $\varepsilon = 0$ and thus $T_s = 0$, but the high reflectivity will cause the plate to appear to have a surface brightness temperature similar to the illumination that it is reflecting. We can thus define a surface scattered radiometric temperature, T_{SC}, to capture this effect, and it can be written as

$$T_{SC} = \rho T_{ILLUMINATOR}, \qquad (7.10)$$

where ρ denotes the reflectivity of the object being illuminated, and the second term on the right-hand side of (7.10) denotes the radiometric temperature of the illuminator. The quantities T_s and T_{SC} can be combined to yield the effective radiometric temperature

$$T_E = T_s + T_{SC} = \varepsilon T_0 + \rho T_{\text{ILLUMINATOR}}. \tag{7.11}$$

Down-welling radiation from the sky is the dominant source of illumination in an outdoor scenario [76]. When a radiometer, designed for the detection of thermal radiation, is pointed toward the zenith, it will detect residual radiation originating from deep space as well as down-wells from the atmosphere. This yields a brightness temperature around 60 K at 94 GHz [71]. A metal object with $\rho = 1$ and $\varepsilon = 0$ will thus have $T_E \approx 60$ K, and as a result it will appear to be very cold in a thermal image. Note however that this scenario is greatly simplified and does not account for many factors that play into the imaging process. An image can thus be formed by measuring T_E as a function of position around the captured scene, generating a 2D map.

7.3.2 Applications

Millimeter-wave imaging—passive or active, in a commercial or military environment have been successfully employed in multiple application areas for a number of years: land vehicle guidance, aircraft landing aids, navigation, security, and many more.

Shoucri et al. [77] developed a 94 GHz passive camera intended for used in aircraft landing assistance systems. The authors highlighted advances in technology that were greatly influential in the development of millimeter-wave imaging systems. The first of these is the advent of high-bandwidth, low-noise millimeter-wave receivers, which greatly enhance dynamic range and, as a result, image contrast. The receiver used was designed for 94 GHz operation with a 8 GHz bandwidth and 5.5 dB noise figure. The worst case integration time per pixel was found to be 2 ms, which corresponds to a temperature resolution of about 0.3 K, comparable to commercial IR cameras available at the time. The second technological advancement that especially influenced real-time image acquisition is focal plane array receivers. With focal plane arrays in 1D or 2D configurations, images could be displayed at up to 30 Hz. This was an enormous improvement over single-receiver systems that needed to mechanically scan across the entire scene to form an image. Finally, newly developed and enhanced signal processing algorithms were influential in improving the optical resolution of a scene, thereby reducing the required size of optics associated with passive sensors.

Clark and Durrant-Whyte [53] and Wehling [78] have investigated the use of millimeter-wave sensors for use in land vehicles. The work done by Wehling focused on implementing multifunction millimeter-wave systems in armored vehicles. According to the author, combining several functions into a single system will be a key factor in reducing cost of millimeter-wave systems. An experimental system operating at 35 GHz was designed that combines active protection radar, surveillance radar, trunking radio for ad hoc networking and combat identification.

7.4   Closing Remarks

Millimeter-wave technology holds immense potential for future systems in communication, security, and safety applications. While significant research is still required in some areas, as we have discussed, the benefits that current millimeter-wave systems provide along with rapidly evolving component technology and fabrication processes have cemented their continued use in the future.

References

1. H.H. Meinel, The current status of millimeter-wave communication systems, in *Mediterranean Electrotechnical Conference* (1989), pp. 680–686
2. H.H. Meinel, Applications of microwaves and millimeterwaves for vehicle communications and control in Europe, in *IEEE MTT-S International Microwave Symposium Digest* (1992), pp. 609–612
3. H.H. Meinel, Millimeter-wave technology advances since 1985 and future trends. IEEE Trans. Microw. Theor. Tech. **39**(5), 759–767 (1991)
4. H.H. Meinel, Commercial applications of millimeterwaves history, present status, and future trends. IEEE Trans. Microw. Theor. Tech. **43**(7), 1639–1653 (1995)
5. J. Proakis, M. Salehi, *Digital Communications*, 4th edn. (McGraw-Hill, New York, 2000)
6. Z. Pi, F. Khan, An introduction to millimeter-wave mobile broadband systems. IEEE Commun. Mag. **49**(6), 101–107 (2011)
7. H. Zirath, T. Masuda, R. Kozhuharov, M. Ferndahl, Development of 60-GHz Ffront-end circuits for a high-data-rate communication system. IEEE J. Solid-State Circuits **39**(10), 1640–1649 (2004)
8. N. Guo, R.C. Qiu, S.S. Mo, K. Takahashi, 60-GHz millimeter-wave radio: principle, technology, and new results. EURASIP J. Wirel. Commun. Netw. **1**, 1–8 (2007)
9. T. Baykas, C.S. Sum, Z. Lan, J. Wang, M.A. Rahman, H. Harada, S. Kato, IEEE 802.15.3c: the first IEEE wireless standard for data rates over 1 Gb/s. IEEE Commun. Mag. **49**(7), 114–121 (2011)
10. M. Di Renzo, H. Haas, A. Ghrayeb, S. Sugiura, L. Hanzo, Spatial modulation for generalized MIMO: challenges, opportunities, and implementation. Proc. IEEE **102**(1), 56–103 (2014)
11. A.L. Swindlehurst, E. Ayanoglu, P. Heydari, F. Capolino, Millimeter-wave massive MIMO: the next wireless revolution? IEEE Commun. Mag. **52**(9), 56–62 (2014)
12. F. Boccardi, R. Heath, A. Lozano, T.L. Marzetta, P. Popovski, Five disruptive technology directions for 5G. IEEE Commun. Mag. **52**(2), 74–80 (2014)
13. A. Natarajan, S.K. Reynolds, M. Tsai, S.T. Nicolson, J.C. Zhan, D.G. Kam, D. Liu, Y.O. Huang, A. Valdes-Garcia, B.A. Floyd, A fully-integrated 16-element phased-array receiver in SiGe BiCMOS for 60-GHz communications. IEEE J. Solid-State Circuits **46**(5), 1059–1075 (2011)
14. A. Siligaris, O. Richard, B. Martineau, C. Mounet, F. Chaix, R. Ferragut, C. Dehos, J. Lanteri, L. Dussopt, S.D. Yamamoto, R. Pilard, P. Busson, A. Cathelin, D. Belot, P. Vincent, A 65-nm

CMOS fully integrated transceiver module for 60-GHz wireless HD applications. IEEE J. Solid-State Circuits **46**(12), 3005–3017 (2011)

15. S. Rangan, T.S. Rappaport, E. Erkip, Millimeter-wave cellular wireless networks: potentials and challenges. Proc. IEEE **102**(3), 366–385 (2014)

16. T.S. Rappaport, R. Mayzus, Y. Azar, K. Wang, G.N. Wong, J.K. Schulz, M. Samimi, F. Gutierrez, Millimeter wave mobile communications for 5G cellular: it will work! IEEE Access **1**, 335–349 (2013)

17. K.S. Rao, G.A. Morin, M.Q. Tang, S. Richard, K.K. Chan, Development of a 45 GHz multiple-beam antenna for military satellite communications. IEEE Trans. Antennas Propag. **43**(10), 1036–1047 (1995)

18. R.K. Crane, Propagation phenomena affecting satellite communication systems operating in the centimeter and millimeter wavelength bands. Proc. IEEE **59**(2), (1971)

19. W.O. Copeland, J.R. Ashwell, G.P. Kefalas, J.C. Wiltse, Millimeter-wave systems applications, in *G-MTT International Microwave Symposium* (1969), pp. 485–488

20. F. Gutierrez, S. Agarwal, K. Parrish, T.S. Rappaport, On-chip integrated antenna structures in CMOS for 60 GHz WPAN systems. IEEE J. Sel. Areas Commun. **27**(8), 1367–1378 (2009)

21. R.C. Daniels, J.N. Murdock, T.S. Rappaport, R.W. Heath, 60 GHz wireless: up close and personal. IEEE Microw. Mag. **11**(7 SUPPL.), (2010)

22. S. Ortiz, The wireless industry begins to embrace femtocells. Computer **41**(7), 14–17 (2008)

23. V. Chandrasekhar, J.G. Andrews, A. Gatherer, Femtocell networks: a survey. IEEE Commun. Mag. **46**(9), 59–67 (2008)

24. S.Y.S. Yeh, S. Talwar, S.L.S. Lee, H.K.H. Kim, WiMAX femtocells: a perspective on network architecture, capacity, and coverage. IEEE Commun. Mag. **46**(10), 58–65 (2008)

25. J.G. Andrews, H. Claussen, M. Dohler, S. Rangan, M.C. Reed, Femtocells: past, present, and future. IEEE J. Sel. Areas Commun. **30**(3), 497–508 (2012)

26. H. Claussen, L.T.W. Ho, L.G. Samuel, Financial analysis of a pico-cellular home network deployment, in *IEEE International Conference on Communications* (2007), pp. 5604–5609

27. M. Abouelseoud, G. Charlton, The effect of human blockage on the performance of millimeter-wave access link for outdoor coverage, in *77th IEEE Vehicular Technology Conference (VTC)* (2013), pp. 1–5

28. T. Bai, R.W. Heath, Analysis of self-body blocking effects in millimeter wave cellular networks, in *48th Asilomar Conference on Signals, Systems and Computers* (2014), pp. 1921–1925

29. A. Brizzi, A. Pellegrini, L. Zhang, Y. Hao, Statistical path-loss model for on-body communications at 94 GHz. IEEE Trans. Antennas Propag. **61**(11), 5744–5753 (2013)

30. D.M. Pozar, *Microwave Engineering*, 4th edn. (Wiley, Hoboken, New Jersey, 2012)

31. Z. Pi, F. Khan, A millimeter-wave massive MIMO system for next generation mobile broadband, in *IEEE Conference on Signals, Systems and Computers* (2012), pp. 693–698

32. J. Hoydis, S. ten Brink, M. Debbah, Massive MIMO: how many antennas do we need? in *49th Annual Conference on Communication, Control, and Computing (Allerton)* (2011), pp. 545–550

33. J. Hoydis, S. Ten Brink, M. Debbah, Massive MIMO in the UL/DL of cellular networks: how many antennas do we need? IEEE J. Sel. Areas Commun. **31**(2), 160–171 (2013)

34. E.G. Larsson, O. Edfors, F. Tufvesson, T.L. Marzetta, Massive MIMO for next generation wireless systems. IEEE Commun. Mag. **52**(2), 186–195 (2014)

35. H. Huh, G. Caire, H.C. Papadopoulos, S.A. Ramprashad, Achieving 'massive MIMO' spectral efficiency with a not-so-large number of antennas. IEEE Trans. Wirel. Commun. **11**(9), 3226–3239 (2012)

36. T.L. Marzetta, Noncooperative cellular wireless with unlimited numbers of base station antennas. IEEE Trans. Wirel. Commun. **9**(11), 3590–3600 (2010)

37. U. Gustavsson, C. Sanchez-Perez, T. Eriksson, F. Athley, G. Durisi, P. Landin, K. Hausmair, C. Fager, L. Svensson, On the impact of hardware impairments on massive MIMO, in *IEEE Globecom Workshop—Massive MIMO: From Theory to Practice* (2014), pp. 294–300

38. H. Ying, D. Xu, B. Ji, Y. Huang, Y. Luxi, Energy-efficiency of very large multi-user MIMO systems, in *International Conference on Wireless Communications and Signal Processing (WCSP)* (2012), pp. 1–5
39. H.Q. Ngo, E.G. Larsson, T.L. Marzetta, Energy and spectral efficiency of very large multiuser MIMO systems. IEEE Trans. Commun. **61**(4), 1436–1449 (2013)
40. A. Hassanien, S.A. Vorobyov, Transmit/receive beamforming for MIMO radar with colocated antennas, in *IEEE International Conference on Acoustics, Speech and Signal Processing*, vol. 1 (2009), pp. 2089–2092
41. A. Pitarokoilis, S.K. Mohammed, E.G. Larsson, On the optimality of single-carrier transmission in large-scale antenna systems. IEEE Wirel. Commun. Lett. **1**(4), 276–279 (2012)
42. L.J. Cimini, Analysis and simulation of a digital mobile channel using orthogonal frequency division multiplexing. Commun. IEEE Trans. **33**(7), 665–675 (1985)
43. S.C. Thompson, A.U. Ahmed, J.G. Proakis, J.R. Zeidler, M.J. Geile, Constant envelope OFDM. IEEE Trans. Commun. **56**(8), 1300–1312 (2008)
44. G.W. Stimson, *Introduction to Airborne Radar*, 2nd edn. (Scitech Publishing, Raleigh, North Carolina, 1998)
45. M.A. Richards, J.A. Scheer, W.A. Holm, *Principles of Modern Radar—Basic Principles* (Scitech Publishing, Edison, New Jersey, 2010)
46. M. Skolnik, *Radar Handbook*, 3rd edn. (McGraw-Hill, New York, 2008)
47. M.A. Richards, *Fundamentals of Radar Signal Processing* (McGraw-Hill, New York, 2005)
48. M.I. Skolnik, *Introduction to Radar Systems* (McGraw-Hill, New York, 1962)
49. A.W. Rihaczek, Radar resolution properties of pulse trains. Proc. IEEE **52**, (1964)
50. C.A. Balanis, *Antenna Theory: Analysis and Design*, 3rd edn. (Wiley, Hoboken, New Jersey, 2005)
51. S.M. Sherman, *Monopulse Principles and Techniques*, 2nd edn. (Artech House Inc, Dedham, Massachussets, 2011)
52. W. Menzel, Millimeter-wave radar for civil applications, in *European Radar Conference (EuRAD)* (2010), pp. 89–92
53. S. Clark, H.F. Durrant-Whyte, Autonomous land vehicle navigation using millimeter wave radar, in *IEEE International Conference on Robotics and Automation* (1998), pp. 3697–3702
54. S. Clark, G. Dissanayake, Simultaneous localisation and map building using millimeter wave radar to extract natural features, in *IEEE International Conference on Robotics and Automation* (1999), pp. 1316–1321
55. F. Sadjadi, M. Helgeson, J. Radke, G. Stein, Radar synthetic vision system for adverse weather aircraft landing. IEEE Trans. Aerosp. Electron. Syst. **35**(1), 2–14 (1999)
56. A. Jain, Applications of millimeter-wave radars to airport surface surveillance, in *13th AIAA/IEEE Digital Avionics Systems Conference (DASC)* (2000), pp. 528–533
57. J. Hasch, E. Topak, R. Schnabel, T. Zwick, R. Weigel, C. Waldschmidt, Millimeter-wave technology for automotive radar sensors in the 77 GHz frequency band. IEEE Trans. Microw. Theor. Tech. **60**(3), 845–860 (2012)
58. T. Takehana, H. Iwamoto, T. Sakamoto, T. Nogami, Millimeter-wave radars for automotive use, in *International Congress on Transportation Electronics* (1988), pp. 131–145
59. F. Pfeiffer, E.M. Biebl, Inductive compensation of high-permittivity coatings on automobile long-range radar radomes. IEEE Trans. Microw. Theor. Tech. **57**(11), 2627–2632 (2009)
60. J. Wenger, Automotive radar—status and perspectives, in *IEEE Compound Semiconductor Integrated Circuit Symposium* (2005), pp. 21–24
61. European Telecommunications Standards Institute, Electromagnetic compatibility (EMC) standard for radio equipment and services. Part 1: common technical requirements. Intellect. Prop. **1**, 1–35 (2002)
62. D.A. Williams, Millimetre wave radars for automotive applications, in *IEEE MTT-S International Microwave Symposium Digest*, vol. 2 (1992), pp. 721–724
63. S. Tokoro, Automotive application systems of a millimeter-wave radar, in *Proceedings of Conference on Intelligent Vehicles* (1996), pp. 260–265

64. L. Raffaelli, Millimeter-wave automotive radars and related technology, in *IEEE MTT-S International Microwave Symposium Digest* (1996), pp. 35–38
65. J. Böck, H. Schäfer, K. Aufinger, R. Stengl, S. Boguth, R. Schreiter, M. Rest, H. Knapp, M. Wurzer, W. Perndl, T. Böttner, T.F. Meister, SiGe bipolar technology for automotive radar applications, in *Proceedings of the 2004 Meeting on Bipolar/BiCMOS Circuits and Technology* (2004), pp. 4–7
66. I. Gresham, N. Jain, T. Budka, A. Alexanian, N. Kinayman, B. Ziegner, S. Brown, P. Staecker, A compact manufacturable 76-77-GHz radar module for commercial ACC applications. IEEE Trans. Microw. Theor. Tech. **49**(1), 44–58 (2001)
67. S. Lutz, D. Ellenrieder, T. Walter, R. Weigel, On fast chirp modulations and compressed sensing for automotive radar applications, in *Proceedings of the International Radar Symposium* (2014)
68. O. Sawade, B. Schaufele, J. Buttgereit, I. Radusch, A cooperative active blind spot assistant as example for next-gen cooperative driver assistance systems (CoDAS), in *IEEE Intelligent Vehicles Symposium* (2014), pp. 76–81
69. M. Dudek, I. Nasr, G. Bozsik, M. Hamouda, D. Kissinger, G. Fischer, System analysis of a phased-array radar applying adaptive beam-control for future automotive safety applications. IEEE Trans. Veh. Technol. **64**(1), 34–47 (2015)
70. D.M. Sheen, D.L. McMakin, T.E. Hall, Three-dimensional millimeter-wave imaging for concealed weapon detection. IEEE Trans. Microw. Theor. Tech. **49**(9), 1581–1592 (2001)
71. L. Yujiri, M. Shoucri, Passive millimeter-wave imaging. IEEE Microw. Mag. **4**(3), 39–50 (2003)
72. B. Kapilevich, B. Litvak, A. Shulzinger, Passive non-imaging mm-wave sensor for detecting hidden objects, in *IEEE International Conference on Microwaves, Communications, Antennas and Electronic Systems* (2013), pp. 1–4
73. B. Kapilevich, B. Litvak, A. Shulzinger, M. Einat, Portable passive millimeter-wave sensor for detecting concealed weapons and explosives hidden on a human body. IEEE Sens. J. **13**(11), 4224–4228 (2013)
74. S.S. Ahmed, A. Genghammer, A. Schiessl, L.-P. Schmidt, Fully electronic E-band personnel imager of 2 m aperture based on a multistatic architecture. IEEE Trans. Microw. Theor. Tech. **61**(1), 651–657 (2013)
75. Z. Xiao, T. Hu, J. Xu, Research on millimeter-wave radiometric imaging for concealed contraband detection on personnel, in *IEEE International Workshop on Imaging Systems and Techniques* (2009), pp. 136–140
76. N.A. Salmon, R. Appleby, Millimetre wave sky radiation temperature fluctuations, in *25th International Conference on Infrared and Millimeter Waves* (2000), pp. 463–464
77. M. Shoucri, R. Davidheiser, B. Hauss, P. Lee, M. Mussetto, S. Young, L. Yujiri, A passive millimeter wave camera for aircraft landing in low visibility conditions. IEEE Aerosp. Electron. Syst. Mag. **10**(5), 37–42 (1994)
78. J.H. Wehling, Multifunction millimeter-wave systems for armored vehicle application. IEEE Trans. Microw. Theor. Tech. **53**(3 II), 1021–1025 (2005)

Printed in the United States
By Bookmasters